Letti Create Your Day

Volume 1

by Paul A. Bartz

The Scripture quotations in this publication are from **The Holy Bible, King James Version,** World Publishing Company.

Letting God Create Your Day, Volume 1 Fourth Edition
by Paul A. Bartz

Copyright © 1991, 2001, 2005, 2008 Creation Moments, Inc.

Creation Moments, Inc.
P.O. Box 839 Foley, Minnesota 56329
creationmoments.com
800-422-4253

ISBN: 1-882510-17-8

All rights reserved. No portion of this book may be reproduced, stored in a retrieval system, or transmitted in any form or by any means, electronic, mechanical, photocopying, recording, or otherwise, without prior written permission from the publisher.

Printed in the United States of America
Printing and production costs for this book were underwritten by friends and supporters of Creation Moments.

Foreword

I am thankful and honored to be able to offer this volume of *Letting God Create Your Day*.

One of the reasons I feel both thankful and honored, as I prepare these scripts and devotions, may not be immediately obvious. Over the past months I have talked to many broadcasters and listeners of "Creation Moments™". Many have contacted us simply to say "thanks" for "Creation Moments™". While doing the Lord's work does not always bring approval, it's always a blessing to have the support of God's people.

I speak on behalf of everyone on the staff of Creation Moments when I tell you that we are all thankful and honored to be able to continue offering "Creation Moments™" to you.

One of our most popular types of program, according to your comments, is a program that presents a creature that is both unusual and surprising. I have often wondered at the universal appeal to this type of program. I have come to the conclusion that I am not the only one who senses a bit of exhilaration when I learn about such creatures. That sense of thrill seems to be fairly universal.

I think that may tell us something about why God made them. His work is filled with what appears to us to be unnecessarily wonderful creatures. That there are so many of them suggests that He made some creatures simply for the joy of creating them. This satisfaction is reflected in Scripture when we are told that God looked at everything He had made and it was very good.

I believe that the exhilaration we sense when we hear about these creatures is at least a shadow of what our Creator felt when He made them.

This joy is like the joy we human beings feel when we do a job well. The sense of a job "well done" makes the extra time and care we put into the effort even more worthwhile. It's more valuable than recognition by others. Yet that recognition, too, is good to hear.

These thoughts should direct our attention to our continued praise and thanksgiving to God for all that He has created. He would be worthy of that praise even if He had not sent His only Son to redeem us from sin, death and the devil. That He was moved by His love for us to carry out His plan of salvation gives us another cause for thanksgiving as well as wonder at His wisdom! Make no mistake; God's forgiving grace is not *just another* cause for thanksgiving. Christ *is* Cause of all causes for our worship.

It is my prayer that this edition of **Letting God Create Your Day** will edify you and draw you closer to our Creator and His saving grace to you in our Lord Jesus Christ.

Pastor Paul A. Bartz
Author, "Creation Moments™"

Fish Learn in Schools, Too!

Psalm 119:73
"Thy hands have made me and fashioned me: give me understanding, that I may learn thy commandments."

Today we are going to a school of fish to learn about the wisdom that God built into some 20,000 different species of fish.

First of all, fish don't learn how to school; they know how to school instinctively at birth. Schooling has advantages for fish, and some schools can have over a million fish in them! You see, schools offer a lot of protection from predators.

Did you ever wonder why, when a school of fish is attacked or simply startled, the fish don't run into each other – but radiate away from the threat like exploding fireworks? Oh yes, they can all see each other; they can tell where the fish next to them are going. But schooling fish also have special organs, called lateral lines, running along the length of their bodies just under the skin. These organs are filled with sensitive hairs, which sense the movement of water around the fish, so that each fish has a better sense of the speed and direction of its neighbor from subtle movements in the water.

It is clear that the advantages of schooling, plus the special organs that schooling fish have to help them school, are not the products of genetic accidents and chance. They are testimony to a Creator Who cares for all the creatures He has made. And if He so cares for fish, how much more does He seek you and desire a relationship with you through His Son, Jesus Christ!

Prayer: Dear Father, You let nothing in Your creation stand and wait for Your love, and You are always with us. Help me to learn more of Your presence in my life and better learn to reject those who account for everything in life by chance or accident. In Jesus' Name. Amen.

REF: Lemonick, Michael D., 1985, "Who do fish fraternize in formation?" *Science Digest*, Apr. p. 88.

Why Are There Germs?

Matthew 13:22
"He also that received seed among the thorns is he that heareth the word; and the care of this world, and the deceitfulness of riches, choke the word, and he becometh unfruitful."

Christians, especially those who believe in creation, hear this question often, "If God created the world perfectly, why are there germs that cause disease? Why did He make them?"

The fact is, only a tiny minority of all microbes cause disease. Almost all microbes, or germs, are "good guys." Not only do they make life possible on Earth, but many of them have been put to work helping humans. The most obvious example is antibiotics, which have saved millions of lives since they came into widespread use in this century. But even 2,500 years ago the Chinese were using antibiotics when they treated boils with moldy soybean curds. And microbes make many other common medicines that save millions of lives.

Bacteria are used to turn plant stems into ethanol, which is used to produce gasohol. Some microbes help recover trapped oil, while others help clean oil spills off beaches. Bacteria mine uranium and other important minerals. Without microbes, we would have no cheese, soy sauce or bread. And microbes have even been put to work producing highly nutritious animal feeds. Without microbes, we would be up to our ears in sewage and undecayed wood.

Microbes are one of our Creator's most important creations. He didn't make any of them harmful to other life. They became harmful as part of the curse upon creation that resulted from sin. We thank God that in Jesus Christ He has provided a remedy for our greatest danger – eternal condemnation because of our sin!

Prayer: Dear Lord, I thank You that You have provided us with so many good microbes and the knowledge to use them. Help me not to focus only on my earthly needs but always to look to Your salvation as my ultimate help. Amen.

Ref: Morgan, D., and T. Monmaney. 1985, "The bug catalog," *Science* 85, July-August, p. 37.

The Loving Poison Dart Frog

1 John 4:19
"We love him, because he first loved us."

The Choco Indians of Panama and Colombia use the poison from the skin of the beautiful poison dart frog to make their lethal darts. The bright orange and deep blue skin of this frog serves to warn predators that it is best left alone and its poisonous skin untouched.

Although it is deadly, the poison dart frog is one of the most loving parents in the entire amphibian world. The female will lay about a dozen eggs in the leaf litter within her mate's territory. Both parents will stand watch over the eggs, keeping them moist, until the tadpoles emerge. Then the female allows each tadpole, one at a time, to wriggle onto her back. She takes each tadpole, in its turn, to its own miniature pond created by water trapped in the fronds of jungle plants. The mother poison dart frog remembers where each one of her tadpoles is and returns on a regular schedule to lay infertile eggs for the growing youngster to eat.

I would prefer to think that the care of the adult poison dart frogs for their children grows out of a sense of love for their offspring, and we know that God is the author of all love. But even if this care is programmed instinct, we must still find the "programmer" – and that takes us back to the Creator once again. Such wisdom cannot be said to come from nowhere.

Prayer: Dear Heavenly Father, the source of all love and wisdom, grant wisdom to Your people so that they may effectively witness to those around them who are being misled to believe that love is merely instinct and that wisdom can come from nowhere. In Jesus' Name. Amen.

REF: Brownlee, Shannon, 1985, *Discover*, May, p. 55.

The Mountain of the Mists

Genesis 18:14
"Is any thing too hard for the Lord? At the appointed time I will return unto thee, according to the time of life, and Sarah shall have a son."

Deep in the Venezuelan jungle, 60 miles from the nearest settlement, it appears as if God decided to plant a very special garden. That garden grows on a mountain that rises 7,000 feet on sheer cliffs above the jungle. The mountain, rising eventually to 9,000 feet, is the highest mountain in South America outside the Andean mountain chain.

Called the Mountain of the Mists, this horseshoe-shaped mass features a 25-mile-long river valley that is just as deep as North America's Grand Canyon. The mountain is so remote and hard to reach that it was virtually unexplored until 1984.

Scientists have found that over half of the species of living things found on the mountain are new to science. Some are unlike any other creatures ever seen on Earth. A plant called the *Neblinaria* resembles a large artichoke standing four feet above the ground on a thick stalk. Four-foot tall carnivorous pitcher plants, giant earthworms, and long-legged frogs add to the uniqueness of the place. Rainfall has washed most of the nutrients from the soil, but many plants make their living by producing their own soil in the center of their leaves, while others live off the air itself.

There is absolutely no limit to God's imagination. Whether the problem to be solved involves inventing creative ways to allow plants to survive, or whether it's a personal problem that seems beyond solution, your Creator is more than willing and able to help you. Seek Him in the Bible today.

> ***Prayer:*** *Dear Lord, it is true that when I am overwhelmed with difficulty, I seem to forget that nothing is too hard for You to solve. Renew my trust in You as I renew my daily commitment to study Your Word. Amen.*

REF: Blonston, G. 1985. "Mountain of the mists." *Science* 85, July-August, p. 61.

The Animal that Confused Scientists

Psalm 106:7
"Our fathers understood not thy wonders in Egypt; they remembered not the multitude of thy mercies; but provoked him at the sea, even at the Red Sea."

Sometimes God's imagination is so creative that even the scientists are stumped by the results. One such example is the colugo, an animal that scientists tried to place in three different families before they finally gave up and designated a whole new family just for that one strange creature.

Scientists agree that the colugo is a mammal. But beyond that, they're stumped. For a while they had classified the colugo as a bat, because it has a membrane that it uses for gliding. But upon closer inspection, it was clear that the colugo was not a bat. The colugo has both front and hind legs and a tail that all are used to stretch the membrane of the cat-sized animal to a width of four feet. Based on this, scientists thought it might be a lemur with wings, so they classified the colugo as a primate. The strictly vegetarian colugo is not a lemur, however. So about 50 years ago, scientists decided to give the colugo its own classification – *Dermoptera*, meaning "skin wing."

Scientists classify animals according to an imagined evolutionary history – the most popular school of thought today. This doesn't prove evolution; it is simply arranging similar things together, just as kindergarten children do as they learn about different objects. But the colugo is such a testimony to God's imaginative creativity that scientists have not been able to find anything that even remotely resembles this strange and wonderful creature.

Prayer: Dear Father in heaven, too often I take for granted the wonderful things You have made. Sometimes it takes something like the colugo to renew my appreciation for what You have done. Help me never to consider the saving work of Your Son Jesus Christ for me without complete thankfulness. In His Name. Amen.

Nature's Shark Repellent

1 Peter 5:8
"Be sober, be vigilant; because your adversary the devil, as a roaring lion, walketh about, seeking whom he may devour:"

Over the centuries humans have tried different strategies to protect themselves from sharks. But it's no secret that shark repellents simply don't work. But now scientists have learned, from a lowly fish no less, that repelling a shark is as simple as washing his mouth out with soap – released from a squirt gun, of course.

The Moses sole, a slow-swimming, apparently defenseless little fish, lives in the Red Sea. What puzzled scientists was that such seemingly easy prey never seemed to be bothered by predators. Researchers wanted to know what the Moses sole's secret was, so they offered one to sharks in a laboratory tank, knowing that sharks will eat anything. Sure enough, the sharks grabbed the sole, but just as quickly they let it go. More study showed that the sole releases a milky, detergent-like substance from the pores in its skin. As soon as the shark got a little of the detergent in its mouth, it wanted nothing more to do with the sole.

While the sole's detergent is poisonous, researchers found that it's the detergent and not the poison that drives sharks away. They have found that a commercially available detergent works even better than the sole's defense.

Just as the shark cannot be tamed, neither can our sinful deeds. This is why God encourages us to stay away from opportunities to sin. He doesn't want us to suffer the necessary consequences of sin.

Prayer: Dear Lord Jesus, I know that sin is wrong because it hurts me and drives me away from You. Create in me a clean heart that has no desire to flirt with sin but seeks to draw ever closer to You, in Whom I am washed clean. Amen.

Ref: *Discover*, July 1985, p. 38.

Feline Secrets

Genesis 2:20
"And Adam gave names to all cattle, and to all fowl of the air, and to every beast of the field; but for Adam there was not found an help meet for him."

A recent survey of pet owners revealed that more Americans now have cats as pets than dogs. While dogs may be "man's best friend," cats have, in certain cultures, been considered the companion of choice.

Scientists have found that the popular belief that a cat's whiskers measure the width of a space it is about to enter is in error. Cats have 25 to 30 long whiskers on their heads. Each whisker is attached to its own nerve in the skin, allowing the cat to pinpoint every twig as it maneuvers through underbrush.

A steady stare between two cats can be understood as a challenge. But a slow blink is a sign of acceptance. If you have a cat, try slowly closing your eyes and opening them again; you may find that your cat will return your signal as a sign of friendship. And did you know that you can use your cat to tell the temperature of a room? Dr. Hans Precht, a German animal expert, found that if a room's temperature is above 70 degrees, a cat prefers to lay in a straight position with head lifted and paws extended. From this point, the cooler the room, the more the cat curls its body and the lower it tucks its head, until, at 55 degrees, the cat is curled into a tight ball with its head against its hind leg.

Thankfully, God knows that we need companionship. After all, He created humans for the very purpose of having companionship with Himself. But we are not like pets, for He spared nothing, not even the life of His Son, to save us from our sin.

Prayer: Dear Father, I thank You for the companionship which pets provide. Help me to value companionship with You above all relationships. In Jesus' Name. Amen.

The Oldest Dinosaur

Romans 1:25
"Who changed the truth of God into a lie, and worshipped and served the creature more than the Creator, who is blessed for ever. Amen."

Many people mistakenly believe that dinosaurs prove evolution. After all, dinosaurs are strange creatures and seem to represent a very different world from the one we know today. But a recent discovery of the most ancient dinosaur challenges this.

The oldest dinosaur ever discovered was about six feet long and weighed about 300 pounds. Scientists had never before seen a complete skeleton of the creature they called *Herrerasaurus*. But being the most ancient dinosaur, perhaps the dinosaur from which all others evolved, evolutionary scientists expected it to be much less advanced that later dinosaurs. What they found was yet another well-designed creature.

The meat-eating dinosaur had excellent teeth, like those of a shark; its jawbone had a double hinge so that it could easily hang onto its prey. These sophisticated features weren't expected in such an ancient dinosaur. This is yet another proof that all creatures were well-designed from the beginning. Herrerasaurus is yet one more example of the impossibility of evolution.

Dinosaurs, just like all other creatures, were created by God during the first six days the universe existed. This means that dinosaurs and humans once shared this Earth in the recent past. All things were created in a perfect state, so there was no need for evolved improvements. The changes we see since creation are not due to evolution but due to the degenerating effects of sin.

Prayer: Dear Lord, teach me to seek my improvement not through created things but through You, the Author of all that is good and the conqueror of all that is evil through Your death on the cross of Calvary and Your Resurrection. Amen.

Ref: "Oldest dinosaur." *Time*, Nov. 13, 1989, p. 75.

Tasteful Data Preprocessing

Job 12: 9-12
"Who knoweth not in all these that the hand of the Lord hath wrought this? In whose hand is the soul of every living thing, and the breath of all mankind? Doth not the ear try words? and the mouth taste his meat? With the ancient is wisdom; and in length of days understanding."

Computer engineers are learning how to speed up computers and make them more efficient by using a number of processors instead of just one. They have also learned that designing computers that process some results and then send those results to another, larger processor for more complicated processing, helps them solve even more complicated problems.

But computer scientists recognize that they were not the first to use this principle. Though these systems are much simpler than the human brain, they mirror the brain's basic structure. God has filled His creatures with biological preprocessors. Even your tongue, scientists have now learned, has preprocessors to help your brain sense a multitude of flavors.

The old theory that taste buds could only sense salt, sour, bitter and sweet has now been replaced by a more complete understanding. Each of your taste buds is made up of about 40 taste-sensing cells that can detect each of the four flavors in various combinations. These cells communicate with each other, in effect deciding what signals to send to the brain about flavor. Your brain then further processes these signals before you learn that the chocolate ice cream you're eating is extra tasty.

Computer scientists have been applying their best knowledge for decades in order to arrive at computers that are this well-designed. This helps illustrate the point that even our tongues were designed by a great intelligence so that we could have the blessing of taste.

Prayer: Dear Father, I thank You for the gift of taste and the ability to enjoy flavors. But do not let me become so centered on my senses that I misuse Your blessings by overeating. In Jesus' Name. Amen.

Ref: Weiss, Rick. 1989. "Taste buds engage in cross-talk." *Science News*, Nov. 11, p. 317.

Why Don't You Rust?

Acts 7:49-50
"Heaven is my throne, and earth is my footstool: what house will ye build me? saith the Lord: or what is the place of my rest? Hath not my hand made all these things?"

Did you ever wonder why you don't rust? Before you laugh, remember – iron is not only a major part of your blood, it is used to attract the oxygen that is carried by your blood to the rest of your body. And you know what happens when oxygen meets the iron in your car or a tool that is left outside overnight – rust! So, why don't we rust?

Part of the reason is that the molecular structure of hemoglobin is very cleverly designed so that the iron in your hemoglobin attracts oxygen and holds it but, at the same time, is prevented from rusting. There are many designs that hemoglobin could have – but the actual structure prevents rust from forming. But 200 billion of your red blood cells die every day. The iron in those cells is no longer prevented by the hemoglobin from forming rust. So your body collects iron from these cells and stores it in tiny protective containers made up of the protein ferritin, where it is prevented from combining with oxygen and turning into rust.

Actually, these rust-proofing systems in the body can go wrong because of a rare genetic defect called bronze anemia. People suffering from this defect actually do develop rust-like deposits in their bodies, and sometimes the rust actually discolors their skin.

The careful planning and intricately related systems that make life possible speak loudly of a Creator. No wonder discoveries in the biological sciences have caused some scientists to defect from evolution and see creation as the better explanation.

Prayer: Dear Lord, everything You have made is excellent beyond compare and made with great understanding. I ask that You would help me to have wisdom from You for my life. Amen.

Ref: "Why don't we rust?" *Science Digest*, May 1984. p. 88.

Wernher von Braun on Creation

Psalm 14:1
"The fool hath said in his heart, There is no God. They are corrupt, they have done abominable works, there is none that doeth good."

Dr. Wernher von Braun was considered one of the 20th century's greatest scientists. After pioneering work in rocketry, von Braun developed the Saturn V rocket, which successfully powered the first manned moon landing. In 1972, this great scientist was asked to comment on the case for design as a scientific theory for the origin of the universe. Aside from the brief comment at the end, the rest is in Dr. von Braun's words:

"For me, the idea of a creation is not conceivable without invoking the necessity of design. One cannot be exposed to the law and order of the universe without concluding that there must be design and purpose... While the admission of a design for the universe ultimately raises the question of a Designer (a subject outside of science), the scientific method does not allow us to exclude data which lead to the conclusion that the universe, life and man are based on design...

"Some people say that science has been unable to prove the existence of a Designer... But they still maintain that since science has provided us with so many answers, the day will soon arrive when we will be able to understand even the creation of the fundamental laws of nature without divine intent. They challenge science to prove the existence of God. But must one really light a candle to see the sun?"

Just remember Dr. von Braun next time you hear someone claim that no real scientists accept the existence of a Creator.

Prayer: Dear Father in heaven, You have made it reality impossible for people to deny You. And yet in our sinfulness we seek to run away from Your loving care. Help me in my life so that I may not, on account of sin, seek anything else but drawing closer to You. In Jesus' Name. Amen.

How to Keep Your Castle Fresh

Genesis 27:27
"And he came near, and kissed him: and he smelled the smell of his raiment, and blessed him, and said, See, the smell of my son is as the smell of a field which the LORD hath blessed:"

If you've ever watched movies that take place in the middle ages, you've probably witnessed scenes set in a huge castle. Some movie makers have tried to be true to life, showing dogs and sometimes even other animals scampering about inside the castle. Now and again Henry VIII or Richard IV tosses a bone onto the floor. In short, those old castles were not too clean, and there were plenty of unpleasant odors around.

People 700 years ago didn't like odors in their homes any more than we do today. And just as you might use air fresheners in your home, they used air fresheners, too. The most common practice was called "strewing herbs." Thyme, lavender, rosemary, mints and other herbs were grown just for spreading on the floors. When they were walked on, the aromatics in their stems were released, freshening the air.

In addition, mint and wormwood were placed in cupboards to repel mice. Many people still sprinkle these herbs among linens when they are to be stored for a while to avoid a musty smell. They will also help keep a hideaway bed fresh smelling during those long periods when it may be folded into a couch.

Few people in our present world know all the ways in which previous generations solved the same problems our modern products are designed to solve. I hope this **"Creation Moment"** has served as a reminder that the Creator's provision extends to every human need – even to something as seemingly unimportant as air fresheners.

Prayer: Dear Father, You know our needs even before we ask. Yet You have commanded us to pray and have promised to hear us. Forgive my poor use of prayer; help me to remember and feel welcome to bring all my needs, even those that seem small, to You in prayer. In Jesus' Name. Amen.

Ref: Brandies, Monica. 1985. "Mother Nature's air fresheners." *Gurney's Gardening News*, Dec.-Jan. p. 26.

Giraffes in Antigravity Suits

Job 39:19
"Hast thou given the horse strength? hast thou clothed his neck with thunder?"

The giraffe has a strong heart to pump blood all the way up to its head and strong arteries to withstand the high blood pressure needed to carry the blood to its head. This is the giraffe's so-called "wonder net," which is a network of blood vessels that helps to stabilize the blood pressure in the giraffe's head even when it raises and lowers its head.

But modern science continues to uncover engineering wonders that enable the giraffe to keep blood flowing evenly to its brain and keep blood from pooling in its legs. Researchers have discovered that giraffes, unlike human beings, have a valve in the jugular vein. But these valves work in the wrong direction to help blood stay in the head. Instead, they close when a giraffe lowers its head, preventing used blood from backing up into the brain.

And how does a giraffe, which stays on its feet all day, keep blood from pooling in its legs? Scientists have found that the skin on a giraffe's legs is very tight-fitting. When a giraffe walks, its muscle movement within that tight skin actually helps pump used blood out of the legs.

If life owed its existence to chance and genetic mistakes, we wouldn't have any giraffes today. But what a wonder of God's design these stately creatures are!

Prayer: Dear Lord, there is nothing too hard for You. Help me to remember the example of the giraffe when life seems filled with too many difficult details. May I be reminded to bring all things to You, for You have promised to hear me. Amen.

Ref: Pedley, T.J. 1987. "How giraffes prevent oedema." *Nature* 329:3, Sept., p. 13.

Social Life Among Lobsters

Exodus 15:11
"Who is like unto thee, O LORD, among the gods? who is like thee, glorious in holiness, fearful in praises, doing wonders?"

For years scientists have thought that lobsters were loners. Intensive study of the lifestyle of lobsters now shows that lobsters have their own very complex, bustling society.

The average lobster spends many of its evening hours checking out newcomers and spying on its neighbors. A lobster will walk to neighboring lobster shelters in rocks or other protective structures and stick its head in each one to see if anyone is home. If the neighbor is out, the spying lobster will inspect his neighbor's home. If someone is at home, the two lobsters will face each other. If the visitor is larger than the one already in the shelter, the larger lobster evicts everyone in the shelter. As soon as everyone is out, the larger lobster allows everyone to return to their home.

Lobsters, it seems, are not hermits but prefer to live socially. They often move to new neighborhoods or to new shelters within their own neighborhoods. And when a newcomer enters a neighborhood, all the residents check him out.

The lobster's two-week-long courtship and mating ritual is especially touching in its apparent affection.

There seems to be no limit to the Creator's ability to make unique creatures and provide them with their own, often very complex, way of life. Such imagination is not produced by accident!

Prayer: Dear Father, there is no creature You have made that You do not care about and have not provided for. Renew my appreciation of Your wonderful work of creation and lead me to praise You every day of my life. In Jesus' Name. Amen.

Ref: Ravven, Wallace. 1987. Lobster Lust: "Don Juans of the Deep." *Discover*, Dec. p. 34.

The Universal Seven-Day Week

Exodus 20:11
"For in six days the LORD made heaven and earth, the sea, and all that in them is, and rested the seventh day: wherefore the LORD blessed the sabbath day, and hallowed it."

According to the Bible, God completed creation in six days and rested on the seventh, thus setting the pattern for humans to follow. If the Bible's account of creation is just a myth, or if the days of Genesis 1 are figurative, why is the very fabric of human existence tied to the seven-day week?

The Hebrew word *shabah* means "seven." It comes from the root word that means full or complete. Not only is this word the source of the word "Sabbath," it is also related to the word seven in several languages, including English. Even the Chinese from ancient times have used the seven-day week. The seventh day of the first lunar month is still known among the Chinese as the "birthday of mankind." In ancient India, Egypt, Rome and Greece, the seven-day week was standard. Exceptions are few and far between.

Science has also learned that one of our most basic biological cycles is tied to the seven-day week. Researchers found measurable impairments in subjects who were denied one out of seven days for rest. They also found that no amount of experimental coaxing could change this built-in seven-day cycle to six or eight days.

Evolutionary researchers admit that they do not know when the seven-day week started. But the origin of the seven-day week is clearly presented in Scripture. The seven-day week shows that the days in Genesis 1 are not make-believe, nor are they figurative – they are the revealed truth of God!

Prayer: Dear Lord, days are another of Your gifts to us to use for our good and Your glory. Help me to show the truth of what the Bible says about days by making each day You give me glorify You. Amen.

The Facts of Human Life

Job 10:10-11
"Hast thou not poured me out as milk, and curdled me like cheese? Thou hast clothed me with skin and flesh, and hast fenced me with bones and sinews."

The infamous *Roe v. Wade* Supreme Court decision that legalized abortion on demand was made in 1973. Now, take a look at what science knows about the miracle God performs as He forms a new human being within a mother's womb.

Even though the fertilized egg is only one cell, it is a unique cell. The fertilized human egg is unlike any other fertilized egg, because it is genetically human. Its 46 chromosomes and 30,000 genes have all that is needed to make an adult human being. Within 18 days, long before most mothers are even aware that they may be carrying a child, the heart starts the first of the 3 billion beats it will make in a lifetime. Only 22 days later, brain waves can be recorded and the unborn child is already moving. By eight weeks these movements will include swimming motions and grasping with fingers that already have unique fingerprints.

At four months the unborn baby is growing at a rate that, if continued, would leave him weighing 14 tons at birth. Studies show that by the 28th to 32nd week, the unborn child is actually able to remember things.

Science knows that even though the child is unborn, it is uniquely human from the day of conception. The womb is the Creator's workshop in which He forms another human being He loves. We pray that the unborn may soon be universally seen as just one more stage of human existence, no less human than infants, adults or the elderly.

Prayer: Dear Father, I pray that the time will soon arrive when men repent of their actions against the unborn. Be their strong defender and use me to speak for and defend those who have no other defenders among men. In Jesus' Name. Amen.

Ref: "Facts of life: Human life begins at fertilization." *National Right to Life News*. Aug. 24, 1989.

The World's Strangest Bird

Genesis 18:14a
"Is any thing too hard for the LORD?"

One could say that the South American Hoatzin is truly a strange bird. It makes a striking sight with its blue skin and red eyes, sporting a four-inch crest of spiky feathers on its head. Young Hoatzin have claws on their wings, allowing them to climb trees just like a monkey. This bird becomes an expert underwater swimmer before it can even fly.

The Hoatzin has one more odd characteristic that scientists find the most puzzling of all. Ninety-five percent of its diet is leaves. It's the only bird known to digest its food the same way cows and other ruminants do. Just like ruminants, the Hoatzin uses bacteria to break down the plant material it eats in a special chamber above its stomach.

How do evolutionists explain this oddity? As one evolutionary scientist who has studied the Hoatzin for decades said, "Hoatzins don't seem to follow the rules of evolution." While he admitted to being creative in being able to come up with evolutionary explanations for creatures, he said that he was never able to arrive at an evolutionary explanation for the Hoatzin's digestive arrangement.

This leads us to glorify God for one more aspect of His creative work. In creating the Hoatzin, He was able to design a creature that defies any explanation humans might try to invent to deny that He is the Creator. Truly the creation declares His handiwork!

Prayer: Dear Father, I thank You that Your work bears witness to Your existence far better than could any human being. Yet You depend on us human beings who know of Your love to us in Christ to communicate the gospel of forgiveness through Christ to others in this world. Enable me to offer a better witness. In Jesus' Name. Amen.

Ref: Cowen, Ron. 1989. "Alimentary, my dear Hoatzin." *Science News*, v. 136, Oct. 21. p. 269.

The Fish that Digs a Well

Job 12:7a,8b,9,10
"But ask now the beasts, and they shall teach thee... and the fishes of the sea shall declare unto thee. Who knoweth not in all these that the hand of the LORD hath wrought this? In whose hand is the soul of every living thing, and the breath of all mankind."

Wells are usually dug by people in order to get water. Why, then, would a fish dig a well?

The well-digger fish is a member of the Jawfish family and lives in the warm coastal waters of southern and southeastern Asia. With its large mouth and extra long jaw bones, the well-digger fish is perfectly equipped to dig holes in the sea bottom using its mouth as a dredging machine. Once its hole is deeper than the length of its body, the well-digger collects small pieces of shell and coral and presses them into the side of its hole. Eventually the entire hole is lined with these hard materials, creating a wall and looking very much like an old-fashioned well. And as any good engineer knows, such a lining helps prevent the sand or mud from collapsing back into the hole.

After finishing his well, the well-digger backs into his hole, tail first, safe from all danger. His hole also provides him with a hiding place from which to surprise his prey.

The well-digger can be given credit neither for the engineering skill necessary to line a hole to prevent collapse nor for the special feature of an extra jaw well-designed to build its well. How could a fish, by chance, think of building a well, no less have the engineering wisdom to line it in order to prevent collapse? The well-digger is an example of how the Creator gives gifts that we don't usually think about, including biological design as well as intelligence.

Prayer: Dear Lord, You have done all things well for all of Your creatures that You may be glorified. Fill me with a desire to do all things before me well so that You may be glorified. Amen.

Ref: Chapman, Goeff. "Weird and wonderful; The well-digger."

Millions of Noses

Psalm 40:5
"Many, O LORD my God, are thy wonderful works which thou hast done, and thy thoughts which are to us-ward: they cannot be reckoned up in order unto thee: if I would declare and speak of them, they are more than can be numbered."

The lobster, with its external shell, is classified as a crustacean and is considered by evolutionists to be among the earlier evolved creatures. But recent study of the habits and abilities of lobsters are showing that these creatures are unexpectedly complex; their abilities are equal to the supposedly more evolved creatures.

The lobster's sense of smell and taste is up to a million times more sensitive than ours. Scents are picked up by hair-like tufts on its appendages. When eating, the lobster carefully samples the scent of each item it eats. The lobster never seems to get bored with a long, leisurely sniff of each morsel. In fact, scientists have found that the nerve cells in the lobster's antennules and walking leg hairs are more specialized than those of any other creature.

Lobsters actually use their sense of smell and abilities to create various chemicals as a means of communication. Both males and females who are ready to mate use scent signals to communicate their availability. They generate still other scent signals to communicate their readiness to move on to each new step in the courtship.

Those of us who believe that the lobster, like all other creatures, was created by God are not surprised to learn that lobsters show much more complexity than evolutionists ever expected. No living thing our Creator has made is either simple or ill-equipped for life.

Prayer: Dear Heavenly Father, as the giver of everything, You have illustrated that Your love and generosity has no limit and that You give great gifts even to the seemingly most lowly. Remind me of this especially when I am tempted to complain about my situation. In Jesus' Name. Amen.

Ref: Ravven, Wallace. 1987. "Lobster Lust: Don Juans of the Deep." *Discover*, Dec. p. 34.

Brain Talk

Psalms 119:100
"I understand more than the ancients, because I keep thy precepts."

There is a huge difference between computers and the human brain. While the computer transmits information using electricity, the brain communicates its information using powerful chemicals. Let's say that you accidentally touch a hot pan on the stove. In less than two-tenths of a second, millions of reactions take place while your brain performs a huge number of operations. As a result, you pull your hand away from the hot pan – very quickly!

Depending on the need, your brain uses many different chemicals for communication within itself and with the nerves that connect your brain to the rest of your body. Over 50 such chemicals have so far been identified. And it has been learned that many of these chemicals work in combination with each other so that over 800 different messages are possible.

Nor does the brain have set paths for messages, like a computer. In fact, it has been compared to a chemical soup rather than the circuits of a computer. In addition, the information paths in your brain change based on experience. If we can use computer language for a moment, the brain writes its own programs doing many complicated tasks of which you and I are never aware.

We know that the computer is the product of careful design, but to argue that the much more complex human brain is an accident of nature makes very little sense!

Prayer: Dear Lord Jesus Christ, more than any other thing which science can study, the human brain speaks most clearly of creation. Cause those who see this to desire to know more about You, and use me to help them find out about the saving gospel. Amen.

Ref: Hammer, Signe. 1986. "How does it work?" *Science Digest*, June. p. 45.

Life in Rock

Jeremiah 32:17
"Ah Lord GOD! behold, thou hast made the heaven and the earth by thy great power and stretched out arm, and there is nothing too hard for thee:"

Live things in Antarctica's Victoria Land arctic desert must be very special indeed. The 2,000-square-mile arctic desert appears to be home to no living things. There are neither soil nor plants. All that seems to exist there is lifeless, windswept, glacier-scoured rock.

On a warm day the desert may warm up to near freezing. During the perpetual darkness of winter temperatures plunge to 158 degrees below zero. Yet we know that there are very few places on Earth for which the Creator has not, in His unimagined wisdom, provided life. Scientists have now learned that there is indeed life in Antarctica's desert.

One-tenth of an inch beneath the surface of the porous sandstone lives what biologists have called a miniature rain forest, because of the various layers of life. A black zone made up mainly of fungi is closest to the surface of the rock. Just below that layer, filaments of fungi and clusters of algae grow between the rock crystals. Several species of bacteria have also been found in this thin zone of life. They gain their food from the minerals in the rock as well as nitrogen released by the aurora australis, the southern lights. Together, these organisms provide a perfect ecological balance to continue life.

When the Creator decides that there shall be life no matter how hostile the conditions, He is able to create not just life but a complete, working ecological balance.

Prayer: Dear Father, there is nothing that is too hard for You. Teach me to rely totally on You for everything in my life so that I am not burdened and carried away from You by the cares of this life. In Jesus' Name. Amen.

Ref: Bartusiak, Marcia. 1983. "Living in rock and lichen it." *Science 83.* p. 74.

How Are Humans Different from Animals?

Genesis 1:26
"And God said, Let us make man in our image, after our likeness: and let them have dominion over the fish of the sea, and over the fowl of the air, and over the cattle, and over all the earth, and over every creeping thing that creepeth upon the earth."

It is not unusual to hear humans being referred to as animals. Many people are insulted by that, but few people know what separates humans from animals.

At one time, people thought humans were separated from the animals by their ability to make and use tools. But we now know that many animals make and use tools. Some people have suggested that our ability to communicate separates us from the animals. But if you spend much time with animals, or are familiar with studies in animal communication, you soon realize that many animals can communicate. Some can even communicate abstract ideas. Nor is the ability to form a culture and pass its knowledge down from generation to generation unique to humans.

As we learn more about the abilities God has given to animals, many people have supposed that these vanishing distinctions between humans and animals mean that humans really are just another member of the animal kingdom.

But there is one thing that sets humans apart from animals, something that science can never discover and/or deny. What makes humans distinct from animals is that humans are morally accountable to God. Animals are not responsible for morality. It is for this reason that we are made differently from the animals. Yes, we have a material body as they do, but we also have a spiritual component to our beings like the angels. In this respect we have as much in common with the angels as we do with animals.

Prayer: Dear Father in heaven, You made me because You want to love me, but by my sin I have pushed You away. For Jesus' sake forgive me, cleanse me in His blood, and give me the peace that can only come through an ongoing, healthy relationship with You. Amen.

The Loving Lobster

Song of Solomon 2:13
"The fig tree putteth forth her green figs, and the vines with the tender grape give a good smell. Arise, my love, my fair one, and come away."

In many ways, lobster interaction seems almost human. Lobsters' mating ritual seems, by human standards, to be especially touching.

The ritual sequence begins when a female who is ready to mate lingers near the shelter of a male. The male responds by fanning his swimmerets, the four pairs of paddles beneath his tail. At this point communication is taking place through chemical signals each is releasing into the water. If he is a clawed lobster, he will show her his claws. After a couple of days of this, she enters his shelter, where more chemical signals are traded. This is followed by a sort of ritualized boxing sequence, where claws become boxing gloves. This goes on for several days as she spends increasingly more time in the male's shelter.

When she is ready to mate, she will lift her claws above the male's head in an action biologists call "knighting." Her body then shrinks as she sheds her shell. At first the shell-less female's body is as soft as a jellyfish, and she cannot even stand up. Mating must take place very tenderly, for the male with his sharp shell could severely damage the unprotected female. After mating, the male will protect the female for up to a week until she grows a new shell.

We tend to recognize the tenderness and affection in the lobsters' mating ritual, despite the fact that these creatures are so different from us, because we both have the same Creator Who is Himself the source of all tenderness and affection.

Prayer: Dear God, I know that You are love. But never let me think that Yours is a love that winks at sin. Rather, help me always to see that Your love's deepest expression is in the suffering and death of Your Son, Jesus Christ, in my place so that forgiven through Him, I could be restored to You. Amen.

Ref: Ravven, Wallace. 1987. "Lobster Lust: Don Juans of the Deep." *Discover*, Dec. p. 34.

Animal Talk

John 1:1-3
"In the beginning was the Word, and the Word was with God, and the Word was God. The same was in the beginning with God. All things were made by him; and without him was not any thing made that was made."

We have often pointed out that the ability to communicate is not what separates humans from animals. In previous programs we have marveled at the range of animal communication and how it reflects the intelligence and genius of the Creator. Researchers in animal communication have been no less amazed as they have now begun to learn that many animals actually have languages and that they can invent new words to communicate new dangers to each other and to their young.

In 1914 a hunter in South Africa was commissioned to exterminate a herd of 140 elephants. He managed to kill all but 20. Those 20 became so skilled at avoiding him that he had to give up the hunt. Even though the park became a preserve in 1930 – with the elephants protected – elephants on the preserve today, four generations since the hunt ended, are unusually wary of humans. Even young elephants are quickly taught to avoid humans.

Ground squirrels have two distinct alarm calls, depending upon what is after them. If a ground squirrel hears the alarm call that warns of a bird of prey, he will dive for the nearest protection. But if the alarm signals a digging predator like a badger, ground squirrels will pass nearby burrows to hide in one that has a back door escape route.

The Bible tells us that all things were made through the Word – the Son of God. Therefore we should not be surprised to learn that many creatures besides humans have the ability to communicate very specific messages to each other.

> ***Prayer:*** *Dear Lord Jesus, I rejoice and praise You for Your part in creation as well as Your love for the creation and for me which moved You to win my forgiveness when man brought sin on the whole creation. Amen.*

Ref: Gould, Carol Grant. 1983. "Out of the mouths of beasts." *Science 83*, Apr. p. 69.

A Surprise Platypus

Genesis 1:31
"And God saw every thing that he had made, and, behold, it was very good. And the evening and the morning were the sixth day."

Europe was introduced to Australia's duckbill platypus in 1798. Because of the difficulties of travel in those days, scientists didn't send a live platypus from Australia to the British Museum in London. They sent only a platypus skin. Scientists in London looked at the duck bill, the beaver tail and the webbed feet of this egg-laying mammal and immediately denounced the creature as a hoax.

Two hundred years later the duckbill platypus continues to amaze scientists. Recently researchers discovered a surprising new ability the platypus uses to find food. It seems that the nerves in the platypus's skin, which relay the sense of touch, are also able to sense electricity. Every time we or any living creature uses a muscle, a tiny electric current is generated. When the shrimp that the platypus eats flick their tails, they generate about 200-millionths to 1,000-millionths of a volt of electricity. That small amount of voltage is enough to enable the platypus to sense and locate lunch.

Modern biological research has also shown another mystery about the platypus. At least it's a mystery for evolutionists. While the platypus is classified as a mammal, it is genetically as different from all other mammals as mammals are from birds. Nor is the platypus genetically like the bird. This leaves the platypus with absolutely no evolutionary history, almost as if it had simply popped into existence.

And that's what the Bible says happened when God created the heavens and the Earth and everything in them during creation week!

Prayer: Dear Father, Your wisdom is so far above even the wisest men that when we rely on our own understanding of even the simplest things, we are easily lost in confusion. This is yet another reason I thank You for the revelation of Your truth, love, and wisdom in Holy Scripture. In Jesus' Name. Amen.

Ref: Horton, Elizabeth. 1986. "The electric-cool platypus." *Science Digest*, June. p. 21.

A Real Sea Dragon

Genesis 1:21
"And God created great whales, and every living creature that moveth, which the waters brought forth abundantly, after their kind, and every winged fowl after his kind: and God saw that it was good."

Because the vast open seas were largely unknown to so many in the ancient world, early explorers were filled with great fear about monsters that might live in unknown regions. After centuries of exploration on and under the sea, we have learned that the seas hold far more beauty than beasts.

One of the most beautiful and unusual creatures in the sea is the sea dragon. Sea dragons are not monsters; the largest members of their family reach a length of only 18 inches. They belong to a family that includes more than 200 types of sea horses and pipefishes. Yet they look completely unlike any other fish. Most of them look like seaweed with eyes and a snout, something like a sea horse's, on one end.

The shapes of sea dragons are the most varied and unusual in the animal world, and their colors are among the brightest. Unlike other fish, they have no scales. They draw tiny marine organisms into their mouths through a trap-door jaw that opens and closes more rapidly than the human eye can follow. The eyes of the sea dragon move independently of each other. And when it's time to start a family, it is the father sea dragon that incubates the fertilized eggs in a pouch in his tail.

Like many other creatures, the sea dragon is such a unique creature that evolutionists admit they have no idea how it evolved. These are the very creatures we'd expect to find if all things were created by God!

Prayer: Dear Lord, as I look about the creation I can see that You love beauty. I thank You for all the beauty and wonder You have created and I ask that You would help me to make my life a beautiful offering to You. Amen.

Ref: Pennisi, Elizabeth. 1985. "Ghosts and Dragons." *Discover*, Nov. p. 80.

"Liquid Air" Mimics God's System

Genesis 2:7
"And the LORD God formed man of the dust of the ground, and breathed into his nostrils the breath of life; and man became a living soul."

Before birth, a developing baby breathes amniotic fluid, which delivers dissolved oxygen to the lungs. When very premature babies are born, they often have difficulty breathing because their lungs lack a substance that enables them to get oxygen from the air. The result is often chronic lung disease or other permanent damage.

But now researchers are testing something that is popularly called "liquid air." This liquid, a perfluorocarbon, can carry oxygen to the immature lungs of the infant who is not yet ready to breathe air, much like the amniotic fluid the child would be breathing had it not been born prematurely. Animal tests suggest that infants who were born after only 20 weeks of gestation could be helped to survive through the use of this "liquid air."

Researchers had to admit that if infants can survive at only 20 weeks, some hard questions are raised about current abortion practices. Yet pro-abortion feeling is so strong in some parts of society that researchers felt they had to disown the idea that they wanted even younger premature babies to survive. The development of "liquid air" is going to make it more difficult for some to support abortion as we now know it.

It has taken humans thousands of years of medical development to come up with an advance that begins to mimic the way an unborn infant gets oxygen. This makes it even more difficult to believe that mindless nature created the womb and much easier to believe that the womb is the creation of a loving and wise Creator.

Prayer: Dear Father, Your love and wisdom is evident throughout creation. I ask that as you continue to grant man progress in solving health problems, You would continue to make it more difficult for people to support abortion. In Jesus' Name. Amen.

Ref: "Liquid air" may help save premature babies. *Minnesota Citizens Concerned for Life Newsletter*, Oct. 1989.

Animal Culture

Psalm 26:7
"That I may publish with the voice of thanksgiving, and tell of all thy wondrous works."

Language is an important part of culture. As the theory of evolution was being developed, evolutionists were quite successful in convincing people that language, education and culture separated humans, whom they described as animals, from other animals. After all, they reasoned, we are evolutionarily much more advanced than animals, and that's why we have language and traditions.

We know from the Bible's description of humans and animals that evolutionists are wrong on all counts. Additional support for this comes as naturalists learn that animals not only have language, but also have a learned culture, which they teach to the next generation.

A study of white-crowned sparrows in the San Francisco Bay area shows that the sparrows' songs from one area to another are very much alike, but they have different dialects. Sparrows from opposite ends of the same town may even have variations in their songs. Researchers have found that young sparrows experiment with a variety of sounds – just as human babies babble. But they learn their song in the dialect of the adults around them. Likewise, each dialect group has its own unique variations on the mating ritual, to which only birds from the same dialect group respond. Dozens of birds are known to have dialects.

We now know that many birds learn their language from their parents and culture, just as do human infants. Language and culture are clearly not the result of our evolutionary progress but the gift of our Creator!

Prayer: Dear Lord, You have given us so many things that we take for granted. Help me never to take for granted Your forgiving grace to me. Help me always to live in highest thanksgiving to You. Amen.

Ref: Gould, Carol Grant, 1983, "Out of the mouths of beasts." *Science 83*, April. p. 69.

The Punctual Bitterroot

Luke 8:14
"And that which fell among thorns are they, which, when they have heard, go forth, and are choked with cares and riches and pleasures of this life, and bring no fruit to perfection."

In 1805, when Lewis and Clark crossed the Continental Divide, they saw North American Indians preparing the large roots of a plant for cooking. Lewis writes that he asked to taste a sample of the root. It was probably with some sense of amusement that the Indians, who cook the root before eating it, handed Lewis the raw root – which immediately upon tasting he named the bitterroot.

The Latin name for the bitterroot means resurrection, for it truly seems as though the bitterroot plant comes back to life from death. The bitterroot grows in rocky areas where the soil is thin and anemic. Under these conditions the plant cannot even support a few leaves at the same time that it has flowers. So the leaves appear early in the spring to make food for storage in the large root. Once enough energy is stored, the leaves wither and flower buds appear.

If the bitterroot developed from another plant, how did it survive in such difficult conditions until it adapted to them? And if its development was only a result of an accidental mutation, how did it develop the precise timing that allows for flowering only after energy has been stored, including the energy from the spent leaves?

The bitterroot is another example that only our Creator can give life and provide for its needs in His often unique way. Even when death came upon the human race because we disobeyed Him, He prepared life that could overcome death through the ransoming death and resurrection of His Son, Jesus Christ!

Prayer: Dear Father, do not let me be so overcome with the cares and worries of this life that I forget about eternal life that You have prepared for me through Your Son, Jesus Christ. In His Name. Amen.

Ref: Keithley, W.E. 1989. Bitterroot. *Creation Research Society Quarterly*, vol. 26, Sept. p. 53.

The Fossils Show Creation

Luke 19:40
"And he answered and said unto them, I tell you that, if these should hold their peace, the stones would immediately cry out."

Evolution says that life began with the simplest forms. It took over a billion years just to evolve algae and another billion years for living things to have more than one cell. It took half a billion years of slow development to generate today's creatures. And evolution says that this story comes from the fossil record.

What most people do not know is that there is no such story in the fossil record. And when not writing textbooks or appearing on television, evolutionary scientists will admit that their story of life cannot be found in the fossil record. According to the fossil record, every major family alive today appears suddenly and fully formed in the Cambrian rocks, which contain the first clear evidences of developed life.

Charles Darwin was aware of this. Believing his own theory to be true, he called this problem a real mystery and wrote that it is probably a valid argument against evolution. Darwin wrote that he expected the problem to be solved as more fossils were discovered. But today, well over a century later, the problem remains and was written about in recent history in the *Scientific American*.

So, Christians should not feel intimidated by the claims of scientists. We Christians have our faith by which we interpret what we see in the world. But the evolutionary story of life and the fossils is nothing more than the interpretation of the world according to evolutionary faith. We agree that far greater faith is required to believe in the revelation of Charles Darwin than to believe the revelation of God.

> **Prayer: Dear Lord, men mock what You have revealed in Your Word and try to intimidate Your people by telling us how ignorant our beliefs are. Give Your people, beginning with me, a strong and bold faith in Your revealed Word. In Jesus' Name. Amen.**

Ref: Marland & Rudwick. 1964. "The great intra-Cambrian ice-age." *Scientific American*, v. 211, August. pp. 28-36.

Creation Makes Better Science

Psalms 111:10
"The fear of the LORD is the beginning of wisdom: a good understanding have all they that do his commandment: his praise endureth for ever."

While the origins debate is basically a religious debate, many people have been taught that evolution is science and creation is religion. But even according to the evolutionist's own rules of science, evolution is not good science. Let's consider the claims of evolution in light of what science knows today.

Life has never been seen to develop from nonliving materials. Yet evolution says it did. Mutations, said by evolutionists to have created all the kinds of living things, have never been seen to produce one creature that was more complex or better able to survive. Then again, the moon has moonquakes, a magnetic field and internal heat – all indications that it is far younger than evolutionists believe. The Cretaceous limestone was produced from sediment in water and extends in one continuous band from Northern Ireland through Europe and Asia to Australia. This suggests that the entire area was beneath the sea all at the same time, yet evolution refuses to accept Noah's Flood as a global and historical event.

If language evolved, why are the most ancient languages the most complex? If religion evolved, why do the most ancient forms of religion worship one God, a Creator, while later forms of the same religions have many gods – who are much more like human beings?

I could list many more examples showing that only the Bible's account of history makes sense in light of what we know today.

Prayer: Dear Lord, the very stones cry out that You are indeed Creator and Almighty God. Help me to add my voice to this witness, and tell others of Your forgiving love in Jesus Christ. In His Name. Amen

Ref: Brown, Walter T. 1984. "The scientific case for creation: 116 categories of evidence." *Bible-Science Newsletter*, June-August.

The Wombat's Backward Pouch

Ephesians 2:10
"For we are his workmanship, created in Christ Jesus unto good works, which God hath before ordained that we should walk in them."

Australia has many animals that are not found anywhere else in the world today. One of the most unique is the wombat, which looks like a small bear with brown fur. The wombat is a burrowing animal.

Like many of Australia's animals, the wombat is a marsupial, having a pouch in which its prematurely born young complete their development. The pouches owned by most marsupials are open at the top toward the mother's head. This works fine for the kangaroo and other marsupials – no one ever saw a kangaroo standing on its head.

But the wombat is a burrowing animal. If its pouch opened toward the mother's head, it would very quickly fill with dirt, which wouldn't do the young wombat inside the pouch any good. So, unlike any other marsupial, the wombat's pouch opens toward the animal's hind legs – it points backward!

The marsupial's pouch is a wondrous enough invention all by itself. But this pouch is a clear case of a specialized intelligent design for a unique situation. If the wombat's backward pouch had been produced by mutations, how would baby wombats have gotten by during the millions of years of trial and error needed to redesign the pouch? It's easy to see the wombat's backward pouch as a humorous hint from the Creator that human theories simply cannot replace the account of creation that He has revealed to us in the Bible!

Prayer: Dear Lord, I thank You that You have so carefully planned every detail of the creation for the benefit of every creature. When things don't seem to be going well in my life, comfort me with the fact that You are still in charge and that You care about the details of my life, too. Amen

Ref: Major, Trevor. 1989. "The backward wombat." *Reasoning from Revelation*, August. p. 3.

God Protects Trees from Insects

Job 42:3
"Who is he that hideth counsel without knowledge? therefore have I uttered that I understood not; things too wonderful for me, which I knew not."

The number of different kinds of insects in the world seems almost endless. There are millions. The larva of many of these insects feed on trees and bushes at least until they develop into adults. One would think, then, that we wouldn't have any trees left, since they are so unfairly outnumbered by such hordes.

But we also know from experience that the creation has been designed with careful and intricate balances. In this case, the many different kinds of trees each produce poisons in order to protect themselves from insects. But there is still balance since insects, too, need to make a living. Not every insecticide produced by a tree is poisonous to every insect. So a tree's insecticides simply *limit* the number of insects that can feed on it.

Some trees have pockets of gluey fluids that, when opened by insect mandibles, glue the insect's mouth closed – a very effective way of keeping something from eating you! But most of the insecticides manufactured by trees involve subtle and complex chemistry. One poison fools the insects' system into thinking that it is an essential amino acid – with fatal results.

Where could such advanced knowledge of chemistry come from? For that matter, could such precise balance – that allows for life but keeps it in control – arise all by itself? As our knowledge of science grows, our appreciation of God's work of creation only increases!

Prayer: Dear Father, while the world says that science is making it more difficult to believe in a Creator, I see how our growing knowledge of the creation offers us the knowledge of even more things to praise You about. Do not let our Christian young people be fooled by the false claims against You. In Jesus' Name. Amen.

Ref: Bernhardt, Peter. 1989. Wily Violets & Underground Orchids: *Revelations of a Botanist.* pp. 7-9.

The Sun Praises Its Creator

Psalms 148:3
"Praise ye him, sun and moon: praise him, all ye stars of light."

In Psalms 148:3 the sun, moon and stars are specifically encouraged along with the rest of creation, to praise God. And truly God, as our Creator and our Savior through Jesus Christ, is worthy of all praise. But perhaps many Christians have been a little too ready to accept this biblical language as figurative.

Based on scientific principles discovered by Isaac Newton, a creationist, a new science has developed today that studies sound waves on the sun. The vibrations on the sun's surface can be measured as the surface of the sun moves. Scientists studying these movements on the seething surface of the sun have learned with the help of computers that the sun actually rings like a bell! They have found that the sun rings over so many frequencies that there are several million resonances going on all at the same time.

Truly the sun is making a glorious sound to the Lord! Scientists have long known that the Earth does the same thing. And we have learned that the moon, too, has its own song. So next time you read in Scripture about the creation – the sun, moon and Earth – praising God, don't dismiss the words as figurative!

And once more, when the scientists finally arrived at some knowledge, they found that the Bible had already been there – for thousands of years. That alone is a great reason for us to join the sun, moon and Earth in praising our wonderful Creator!

Prayer: Lord, You are worthy of all praise. Take my life and let it be a living song of praise to You for my salvation as well as all my earthly needs. In Jesus' Name. Amen.

A Disastrous Evolutionary Explanation

Psalms 119:9
"Wherewithal shall a young man cleanse his way? by taking heed thereto according to thy word."

We hear it often: Christian values are to blame for most of the problems in society. Marriage is ridiculed as "just a piece of paper." Strongly held convictions of right and wrong are said to be to blame for arguments, fights and wars. And by all means, we must free ourselves from old-fashioned attitudes about sex and marriage.

Scientific support for these "liberated" values goes back to 1928 when Margaret Mead spent nine months in Samoa, then wrote a book. In it she said that Samoans live a happy, carefree life. Not only is crime virtually unheard of, but even young people pass through their teen years without becoming rebellious. Mead, writing as an anthropologist, said this was because Samoans practice complete sexual freedom and avoid having strong convictions about religious and philosophical matters. For more than 50 years Mead's writings have been cited as scientific support for anti-Christian values.

But now an anthropologist who has spent many years in Samoa has come forward with a book that says things in Samoa are almost exactly the opposite of what Mead reported. The Samoans themselves asked him to correct the record, because almost nothing Mead wrote about their values was accurate. Samoans are very puritanical about sex, and parents are strict with their children and competitive with each other.

How foolish humans are to think they can invent better moral values that produce more happiness than those values their Creator gave them. And what a shame that so many have based their lives on a lie because they wanted to believe that "nature" made them.

Prayer: Dear Heavenly Father, I know that You have taught us what is right and wrong because the wrong is self-destructive. Help me not to trust my own morality for my justification, but trust only in the atonement of Jesus Christ for my sins. In His Name. Amen.

Ref: Rensberger, Boyce. 1983. Margaret Mead. *Science 83*, April. p. 28.

The Gift of Words

Acts 17:26
"And hath made of one blood all nations of men for to dwell on all the face of the earth, and hath determined the times before appointed, and the bounds of their habitation;"

Is language learned or is it a built-in part of our very nature? If we were made by a Creator who values communication, we would expect to have some built-in language skills and desires. In fact, common features among all languages would also support the Bible's historical account of how humans once had only one language.

Researchers who studied human methods of communication around the world learned that all people have far more in common with one another than can be explained by simple biology. Linguists have noted that all languages have verbs, nouns and similar phrase structures. Certain consonants in English serve as stop sounds for words. When researchers tested African infants who had never heard English, they found that those infants recognized the same boundaries as American infants.

Not only is language universal to humans, but also every human language follows the same basic rules, even though this is unnecessary. Some scientists are now beginning to theorize that present human languages are all built on a single, universal language pattern.

They have also noted one more important fact. While you can coax an ape to learn a limited vocabulary by bribing it with bananas and candy, human children happily learn 150 new words per week simply for the joy of learning language! No biological theory of origins can explain this rapid language acquisition. The Bible's history of humanity offers the only possible explanation for human languages.

Prayer: Lord, help me never to abuse the gift of language and words by lying or purposefully confusing others. Help me to thankfully use Your gift of language by communicating the gospel to those around me. In Jesus' Name. Amen.

Ref: Gould, Carol Grant. 1983. "Out of the mouths of beasts." *Science 83*, April. p. 69.

A Real Bat Computer

Psalms 51:12
"Restore unto me the joy of thy salvation; and uphold me with thy free spirit."

Nearly everyone is familiar with the way in which the whistle of a train moving toward them lowers in pitch as the train passes. The change in pitch is caused by the change in the movement of the train relative to the listener – first it is moving toward the listener, but after it passes it is moving away. This change in pitch is called Doppler shifting. Now let's apply this principal to the bat's echolocation system.

Bats are most sensitive to certain frequencies as they listen for the echo of their high-pitched squeak. If you are a bat, you listen for echoes from stationary objects around you, like trees, as well as moving objects, such as an insect that is about to become lunch. Because of the differences in the movements of these objects relative to your flight, Doppler shifting changes the pitch of returning echoes. That change could place a returning echo outside the range of frequencies to which you are most sensitive.

Scientists have discovered that the bat solves this problem by calculating the expected change in frequency due to Doppler shifting and then altering his squeak so that the returning echo is at the needed frequency! How many of us could do that without a computer and other sophisticated equipment?

Amazingly, the Creator has gone so far as to make the bat a bit of a physicist so it can make its living. What an elegant testimony to God's wisdom and generosity!

Prayer: Dear Father, I thank You that in Your generosity You withhold nothing from me that is good and in Your wisdom You don't give me everything that I want. Amen

Ref: "Bats alter frequency of squeaks to aid hunt for prey." *The San Diego Union*, Mon., Oct. 23, 1989. p. D-1.

The Mystery of the Frogfish

Genesis 1:24
"And God said, Let the earth bring forth the living creature after his kind, cattle, and creeping thing, and beast of the earth after his kind: and it was so."

The frogfish is yet another strange creature for which those who believe in evolution can find no history. Not only do evolutionists admit that frogfish are not related by evolution to any other creature, but also they can find no relationship among 35 of the 41 known species of frogfish!

Frogfish live in many different shallow sea areas. They are called frogfish because they have a wide mouth and face like a frog. In addition, their front fins are on the ends of what appear to be very fat frog legs. Yet they also have a fleshy lure, suspended on sort of a pole coming out of their forehead, like the deep-water anglerfish. And like the angler, the frogfish uses this bait to lure lunch close to its mouth.

Like the platypus, the frogfish unites very different characteristics of several creatures into one unlikely but imaginative creation. Evolutionists are unable to imagine an evolutionary history for this creature. What's more, they cannot figure out how most of the 41 different species of frogfish they have designated might be related to one another.

The frogfish is a dramatic testimony to the fact that God has created all creatures according to His unlimited creativity, fully formed and so unique that sometimes evolutionists cannot even invent an evolutionary history for them.

Prayer: Dear Lord, nothing is impossible for You. Remind me of this not only when I look at the world You have created, but also when I study Your Word and when I face problems in my life. In Jesus' Name. Amen.

Evolution's Impact on Society

Romans 2:14-15
"For when the Gentiles, which have not the law, do by nature the things contained in the law, these, having not the law, are a law unto themselves: Which shew the work of the law written in their hearts, their conscience also bearing witness, and their thoughts the mean while accusing or else excusing one another;"

Sociologists are treading more boldly on ground that once was the sacred domain of Christianity. In recent years sociologists have tried to explain why certain universals exist among human beings. They are puzzled, for example, by the fact that a smile or a look of grief looks the same in every culture in the world.

Sociologists admit that they can find no way to explain how evolution could have created these universals. But they are not about to admit that perhaps their problem is that evolution is a myth.

Despite their inability to explain these basic features about people, sociologists are talking ever more seriously about social engineering – changing society according to what they feel are proper principles for evolved creatures. Harvard's Edward Wilson warns that if society continues to live by conscience and what it considers to be God's will, there is little chance of creating what he views as a benevolent world. In other words, belief in God and doing His will among other humans is dangerous to society.

This shows us that what people believe about where they came from isn't simply a dry philosophical question. History is filled with people who used evolution to justify their attempts to do away with God's natural order among human beings. And every one has been a failure.

Prayer: Dear Father in heaven, preserve us from those who would set themselves over us because they think that they can invent a better order among human beings than You have given us. In Jesus' Name. Amen.

Ref: Rensberger, Boyce. 1983. "On becoming human." *Science 83*, April. p. 38.

To Bee or Not to Bee

Isaiah 45:18
"For thus saith the LORD that created the heavens; God himself that formed the earth and made it; he hath established it, he created it not in vain, he formed it to be inhabited: I am the LORD; and there is none else."

They are social creatures that live in a colony. Young workers in the colony gather food and keep the colony's living area clean. Older males breed. And the colony is ruled by a single female – a queen.

Are we talking about ants? Or maybe bees? No, this social creature is a mammal called the naked mole rat. They are found in Kenya. And as one naturalist put it, "There's hardly anything about the animal that isn't astonishing."

A naked mole rat colony can have over 80 members. The largest and most important member of the colony is the queen. As the queen patrols the tunnels of the underground colony, her subjects lower themselves to allow her to pass over them. Should she encounter another female, the subject is required to show honor to her queen by giving her a kiss. None of the other females in the colony are able to reproduce. But should something happen to the queen, other females will start developing the ability to reproduce and will fight each other to the death until one remains as the new queen.

The similarity between these mammals and certain insect colonies offers yet another hint of the fact that one Creator stands over all of the creation. The naked mole rat seems to be a rather humorous reminder by our Creator that natural laws cannot explain where we and everything else came from.

Prayer: Dear Lord, there is no limit to Your imagination and power: Teach me, through Your revealed Word to be more like You and to value the same things You value. In Jesus' Name. Amen.

Ref: "The queen of the molehill." *Science 84*, April 1984. p. 8.

Jet Planes that Taste Bad

Isaiah 45:12
"I have made the earth, and created man upon it. I, even my hands, have stretched out the heavens, and all their host have I commanded."

Many airports around the world have a problem with birds interfering with takeoffs and landings. In 1985 birds caused over half a million dollars worth of damage to airplanes as they collided with the moving planes. But Japan's All Nippon Airways came up with a way to keep birds away from its planes.

All Nippon Airways did this by giving the birds a message: "Our airplanes taste bad." You see, many insects that taste bad to birds have a large spot on them, called an eyespot. When birds see an insect with an eyespot, they avoid it because they know it tastes bad.

Balloons with eyespots painted on them have been successfully used in Japan as scarecrows to keep birds out of crops. Flags with eyespots have been used to keep birds off buildings. So All Nippon Airways painted eyespots on the rotating fans of their jet engines. The spots are painted just a little off center so that as the fan spins, it looks like a very angry eyespot. As a result, birds have been avoiding All Nippon planes. In the year after the eyespots were introduced, only one bird struck a plane.

In our modern age of high technology, it seems strange that we would learn how to solve some of our technological problems from nature. But it really shouldn't seem strange. After all, the Creator who is responsible for the creation knows far more than even the best scientist!

Prayer: Dear Father, teach me and help me to understand the things You would teach me. Let me be immersed in Your Word so that I am instructed by You in preparation for life eternal with You. In Jesus' Name. Amen.

Ref: Boxer, Sarah, ed. 1987. "These eyes high in the Japanese sky are strictly for the birds." *Discover*, January. p. 8.

Why Does the Sun Shine?

I Kings 8:23
"And he said, LORD God of Israel, there is no God like thee, in heaven above, or on earth beneath, who keepest covenant and mercy with thy servants that walk before thee with all their heart:"

If we could harness all of the sun's energy that reaches the Earth, we would have more than enough energy to fill all our energy needs – even if the entire world became as industrialized as the west. Yet less than .10% of all of the sun's energy output reaches the Earth! What is going on within the sun to create so much energy?

Most people have heard that the sun produces its energy through nuclear fusion. And even though most scientists think this is true, they admit that the real answer isn't that simple – and the full answer is unknown. If the sun produces its energy by nuclear fusion, one of the results of that process is a tiny particle called a neutrino. But, in fact, the neutrinos that should be coming from the sun are not. Worse, new calculations of the temperature needed for the sun to ignite into the nuclear fusion process show that the evolutionary theory for how the sun formed does not generate enough heat for fusion.

An alternate theory says that the sun's heat is a result of its powerful gravity pulling the sun in on itself – called gravitational collapse. This would explain why the neutrinos are missing. But evolutionists have rejected gravitational collapse, because if this is where the sun's heat comes from, the sun could not be billions of years old.

Despite our scientific sophistication, many of God's works still defy our understanding!

Prayer: Father, there is no God like You! I thank You that even though You command the stars and all creation, yet You come to me with Your love, inviting me into a personal relationship with You through Your Son, Jesus Christ. In His Name. Amen.

Ref: DeYoung, Don B., and David E. Rush. 1989. "Is the sun an age indicator?" *Creation Research Society, Quarterly,* v. 26, September. p.49.

A 2,000-Year-Old Computer

Genesis 4:21
"And his brother's name was Jubal: he was the father of all such as handle the harp and organ."

According to the Bible's history, humans are – by nature, not by evolution – clever and inventive. In Genesis 4 we read that Jubal, of the eighth generation after Adam, was a master musician. His brother Tubal-Cain was an instructor in metallurgy. Clearly these people were not any less intelligent than we are and, considering what they did, perhaps in some ways they were even smarter than we are.

Now and again evidence of high human intelligence and creativity in ancient times is uncovered by science. In one such instance, a formless lump was discovered in 1900 among some art objects at the site of an ancient shipwreck. As the lump dried out, it split open to reveal bronze plates, gears and engraved scales with first-century BC Greek lettering. It took a great deal of study to assemble the parts into a working device.

What was this strange object? It turned out to be a computing machine which, when the appropriate dials were set, computed the motions of the moon, sun and probably the planets as well. In doing this, the computer used a very accurate Babylonian mathematical method that is not yet fully understood. But following the instructions on the plates and dials, this first-century BC computer will give accurate information on astronomical events for any time period!

This fascinating computer is exactly the sort of thing one would expect from human beings at any point in their history, because God has given us the gift of intelligence, curiosity and creativity!

Prayer: Dear Lord, I thank You for the gift of curiosity and intelligence. Help me to use that gift to Your glory. In Jesus' Name. Amen.

Ref: *Scientific American*, 6/59, pp. 60-67, Derek J. de Solla Price, "An Ancient Greek Computer."

Katydids with Personal Guards

Job 10:12
"Thou hast granted me life and favour, and thy visitation hath preserved my spirit."

Relationships among creatures who are often very different from one another are so numerous – and so intelligent – that they witness to an overall design for living things. Even evolutionists recognize this when they talk about the "grand design" of the living world. And if design is that evident, there must be a designer. But there's even more to the matter than seeing a designer.

Some of the most ferocious wasps in the world live in the tropical forests of Costa Rica. These wasps are social, building a huge nest that can be two feet long and one foot across. They are short-tempered and highly protective of the tree in which they build their nest. Most other creatures give these wasps and their tree plenty of room.

But one other creature also calls the tree home – the katydid. These long-horned grasshoppers spend much of their day sitting in the wasps' tree. The wasps don't bother the katydids, because they offer no threat to the wasps. At the same time, none of the katydids' many predators dare come close to the tree because of the wasps. In one research project, scientists removed the wasp nest from a tree that was home to many katydids. A few days later the katydids were gone from the tree as well.

The relationship between these wasps and the katydids is called symbiosis and shows more than intelligence and design in the creation. It shows that the Creator cares about everything He has made – reminding us that He desires a loving relationship with each of us!

Prayer: Dear Heavenly Father, how often I forget that You care for me and invite me to be even closer to You than I am now. Forgive me and draw me closer to You through Your inviting Word in the Bible. In Jesus' Name. Amen.

Ref: *Science 84*, April 1984, p. 65.

Mice Who Farm

Psalm 145:15
"The eyes of all wait on thee, and thou givest them their meat in due season."

Many different creatures, from mice to insects, eat seeds. The tropical forest offers the greatest variety of seeds found anywhere in the world. The problem is that among this great variety are seeds that are poisonous and others that are hard as rocks. Yet even these seeds are used for food by some creatures.

One vine-like legume that grows in the jungle produces seeds that are highly poisonous to insects. Yet there is one beetle that is able to eat these highly nutritious seeds – even gaining extra nutrition for itself from the poison.

The Hymenaea protects its seeds another way. Hymenaea seeds are locked within a very tough, woody pod that no one in the tropical forest is interested in eating – if they can even get it open. But one mouse that lives on the jungle floor is up to the task. This small mouse opens the pods and stores the seed in its pouch-like cheeks. Back in its nest the mouse bites a small notch in the hard seed to help it absorb water. He then buries it in his den. Before long the seed germinates and the mouse has a tasty and tender seedling to eat.

Evolutionists tell us that farming was one of the major advances that helped primitive humans become modern. How, then, do they explain mice who not only farm but who know how to improve the germination of their crops? A better explanation is that the Creator shared this bit of farming wisdom with the mouse when He made it.

> **Prayer: Dear Lord, I thank You that You have stocked the creation with so much food and that You have given each of Your creatures the wisdom necessary to use that food for their benefit. In Jesus' Name. Amen.**

REF: *Science 84*, April 1984, p. 65.

Did Job Have a Weather Satellite?

Job 37:9
"Out of the south cometh the whirlwind: and cold out of the north."

One of the most amazing things in our modern age is the scientist who thinks he can use science to judge the Bible. After all, many things accepted today as scientific fact were first taught in the Bible.

Job 36:27-28 explains the water cycle in which, through evaporation, tomorrow's rains are drawn into clouds. Ecclesiastes 1:7 explains why the rivers do not fill the seas. It tells us that there is a cycle of water from rivers to seas back to fill the rivers again. It was not until 350 B.C., long after Job was written and more than 600 years after Ecclesiastes was written, that Aristotle began to understand the water cycle. And finally in 1841 a scientist, using a thermometer that Galileo invented in 1593 and a barometer that had been invented by Torricelli in 1643, showed that clouds were actually the result of rising water vapor.

Job 37:9 and Ecclesiastes 1:6 both speak of wind and weather patterns that were finally confirmed in 1940. Read these passages before you look at the latest satellite weather pictures – the satellite clearly shows what Scripture is talking about in these verses.

Many of the accepted facts of today's science were originally stated by God in the Bible. Science should not pass judgment on the Bible. After all, it has taken science thousands of years to begin to catch up with the Bible's level of knowledge about even such a simple thing as the weather.

Prayer: Dear Father, man is a prideful creature who typically thinks he knows more than he really does. Help me to see pride in my life for human pride always stands in the way of a closer relationship with You. In Jesus' Name. Amen.

Natural Insect Repellent

Genesis 1:11
"And God said, Let the earth bring forth grass, the herb yielding seed, and the fruit tree yielding fruit after his kind, whose seed is in itself, upon the earth: and it was so."

One interesting way in which to help young people gain an appreciation for God's creation is to ask them to solve certain problems that the Creator Himself had to solve.

Let's use the examples of seeds. Plants need seeds to produce the next generation. But what about all the insects that feed on seeds? As Creator, one could decide that insects will not eat seeds. But what will insects eat? Can we get rid of insects? If we do, what will we create to do all the important jobs that insects perform in the ecology? The conclusion must be that we need insects, and they need to eat something. Seeds are a convenient food for them. So how do we design seeds so that both plants and insects can benefit?

Left to their own to work out this problem, perhaps in small groups, young people are likely to propose two popular solutions. A plant can be designed to make so many seeds that insects could not possibly eat them all. Or some plants could be designed that produce seeds that have a natural insect repellent.

And both of these solutions to the problem are commonly found among seeds. Studies show that plants that produce so many seeds that insects could not possibly eat them all generally do not produce seeds that are poisonous to insects. But plants that produce fewer and larger seeds often produce seeds laced with natural insect repellents. As a result, insects get to eat and plants are allowed to produce the next generation.

> **Prayer: Dear Father in heaven, I don't always appreciate everything You have made often because I don't know what Your purposes are in making them. Help me to see Your purposes in my life in the faith that Your will is always good for me. In Jesus' Name. Amen.**

Ref: *Science 84*, April 1984, p. 65.

A 2,000-Year-Old Electric Battery

Genesis 4:22a
"And Zillah, she also bare Tubal-cain, an instructor of every artificer in brass and iron:"

While evolutionary historians refer to the ages of humanity before the birth of Christ with terms like the Stone, Bronze and Iron ages, it seems that human metal-working technology was much more sophisticated than these names suggest. And that fits with what Genesis 4 tells us about human history.

An example of this was reported in the April 1957 issue of *Science Digest*. In the 1930s archaeologists were digging in a small ruin on the outskirts of Baghdad when they uncovered something that looked very much like a modern dry-cell battery. The object was analyzed, and a model was built by an engineer at General Electric's High Voltage Laboratory.

The device turned out to be a wet-cell battery that packed enough power to gold plate jewelry – history's first recorded industrial use of electricity. The battery was built and used by Baghdad silversmiths from 250-224 B.C. It was about the size of two modern flashlight batteries and was made of materials similar to those in use today and common in the ancient world, including copper and 60/40 tin-lead alloy solder.

These discoveries make it nearly impossible to dismiss the early chapters of Genesis as mythical. The picture they present of a creative and intelligent humanity rings true. And the Bible's message of humanity's need for salvation in Jesus Christ rings true as well.

Prayer: Dear Lord, in trying to escape from You, man pictures himself as a well advanced animal and nothing more. I am surrounded by people who think that this is scientific fact. Help me to be a witness to the truth revealed in Your Word so that those who have been misled may be led to Christ. In His Name. Amen.

Eating Pollution

Psalm 119:99
"I have more understanding than all my teachers: for thy testimonies are my meditation."

How would you like to live in a neighborhood where garbage was never collected and taken away? What if there was no provision in the creation for getting rid of waste? By now the Earth would have become a filthy mess!

Our sewage and other waste does not break down all by itself. Without microbes, waste would simply continue to pile up. But if you were designing a world in which waste would have to be "taken away," how would you do it? It's hard to imagine a solution to the problem that is better than the solution God provided. The creatures that dispose of our waste are microscopic – an improvement over our big, noisy garbage trucks. But microbes don't simply break down garbage into harmless materials. In His wisdom, God designed them to turn useless garbage into useful substances.

We are now beginning to make more use of what God has designed. A West German corporation now markets a ceramic sponge that is home to countless organisms that not only love highly polluted water, they turn the pollution into useful things like natural gas and alcohol. Scientists say that the bugs eat almost anything, even pollutants that cannot easily be handled by usual sewage treatment.

Whether we recognize it or not, we humans continue to learn at the feet of the Creator who made us – which is, itself, an argument against evolution.

> ***Prayer: Dear Lord, I praise You for the balances You have built into the creation in order to make life livable. Help us to see that though most scientists would deny that You are Creator, Your workmanship in the creation is a witness to You. In Jesus' Name. Amen.***

Ref: "Minute diners love our waste." *The Province*, July, 12, 1989.

The Three-Eyed Porpoise

Psalms 147:5
"Great is our Lord, and of great power: his understanding is infinite."

What has three eyes, lives in the sea and can swim faster than most boats can travel? No, it's not a creature from another planet. It's the porpoise.

Science has long known that the porpoise navigates using sonar. Now they have learned a little bit more about how well equipped the porpoise is in its use of sound waves. Scientists have found that the porpoise has a fatty, disc-shaped organ in the middle of its forehead that is able to shape and focus the sound waves used in the porpoise's sonar. When the porpoise generates sound waves, they are conducted by bones in the mammal's forehead to this disc, which then changes shape to focus and aim the sound waves. The returning echo allows the porpoise to literally see an image.

Scientists note that the porpoise's brain is very highly developed so that it can carry out this complex form of navigation. Because of the nature of sound waves, scientists suspect that a porpoise could not only see a fish but very likely make out some details of its internal organs.

The porpoise's sonar system uses a highly sophisticated knowledge of physics, biology and data processing. No scientist can even begin to explain how mutations and chance could have designed this complex and technically impressive navigation system. But there is no mystery if we understand that the porpoise was designed and made by the Creator of all that science studies!

Prayer: Dear Lord, I marvel and praise You for the excellence of all that You have created. I ask that as our knowledge of the creation grows more people would be led to see and glorify You as Creator, and so be drawn to Your Son, our Savior, Jesus Christ. In His Name. Amen.

Ref: Schwarz, Joel. 1984. "Porpoise secrets." *Omni*, v. 6:6, March. p. 37.

100-Foot Ferns

Genesis 1:31
"And God saw every thing that he had made, and behold it was very good. And the evening and the morning were the sixth day."

Charles Darwin's *On the Origin of Species* opened the modern era of evolutionary thinking. The very name of his book suggests that as we look back in time, we should see fewer kinds of animals. Darwin led many people to think that there should be a greater variety of plants and animals today than in the distant past. And, those plants and animals should be more sophisticated today.

But the fossil record tells just the opposite story. The most ancient rocks with identifiable fossils of multi-celled creatures in them contain every major family alive today. These creatures just appear fully-formed all at once. And both evolutionists and creationists admit that not all the creatures that existed in the day when these fossils were formed have been preserved in the fossil record.

When you think of some of the pictures of strange creatures from the early days of Earth history, it seems obvious that, contrary to Darwin's theory, there was a much greater variety of creatures alive in the past than exists today. And generally they were bigger and stronger than creatures are today. Ferns grew over 100 feet high, and dragonflies had wingspans of six feet. There are fewer kinds of creatures today, and what we have are often smaller and weaker than what we find in the fossil record.

These facts don't present a picture of evolutionary development and improvement at all. It is a picture of a perfect creation corrupted by sin and running down. This is the same flow of history revealed in Scripture. But thankfully, Scripture shows us how in Christ Jesus there is escape from all the effects of sin!

Prayer: Dear Father, the entire creation groans under the consequences of man's sin. Give me a clearer understanding of the fact that the gospel of forgiveness is a needed message for us in the everyday world, and help me to communicate that to others. In Jesus' Name. Amen.

The Brain Repairs Itself

I Kings: 4:29
"And God gave Solomon wisdom and understanding exceedingly much, and largeness of heart, even as the sand that is on the seashore."

One of the most studied but least understood organs in the human body is the brain. The human brain is the most complex arrangement of matter in the universe. There are many different theories about how the brain works, but one universally recognized fact is that the brain is more sensitive to damage than any other organ.

Scientists studying the brain have learned that when the brain is injured, it triggers the release of some very special chemicals called neuronotrophic factors. While researchers aren't sure how these chemicals actually work, studies show that the chemicals seem to help the brain repair damage to itself. They believe that each of the different chemicals works to repair a specific structure in the brain.

Medical researchers are continuing their study of these factors with the hope of using artificially created versions to help those suffering from brain damage due to stroke and trauma. They hope to learn from the body how to help the body do what it now tries to do for itself.

Surely the advanced knowledge of biochemistry necessary to create a way for the brain to repair itself does not come from matter itself. Only a powerful and wise Creator – the Creator described in the Bible – could have made this system. But we speak here of more than His wisdom. For His gift of the brain's ability to repair damage to itself also show that He looks upon us with love. The highest expression of His love to us is found in Jesus Christ.

Prayer: Dear Father, when I am tempted to escape Your anger over sin, remind me that Your deepest desire is that I cling to what Your Son, my Lord and Savior Jesus Christ, has done for my forgiveness. Amen.

Ref: Sobel, David. 1983. "Brain self-repair." *Omni*, v. 5-5, February. p. 39.

An Airplane from Ancient Egypt

Psalm 55: 6-7
"And I said, Oh, that I had wings like a dove! for then I would fly away, and be at rest. Lo, then I would wander far off, and remain in the wilderness."

These days we don't have to wait very long before we hear some nonsense about how flight or – more recently – space travel are but stages in human evolutionary development. As Christians we know that when Adam, the first man, is described as perfect, he is perfect in every way – spiritually, morally and intellectually. This probably explains why Scripture says that within the first few generations of human existence on Earth the arts and sciences were flourishing.

In 1898 an object that was thought to be a small wooden model of a bird was discovered in Egypt. The "unremarkable" find was tossed into a box marked "wooden bird models" and was rediscovered in more recent times. The Egyptian Ministry of Culture set up a special panel to study the object. As a result of that study, the object is on display at the Cairo Museum today as the centerpiece of a special exhibit. It is now officially labeled as a model airplane.

The plane's design shows that it is an excellent glider, and with a small engine it could carry huge loads because of the unique design of its wings. This 2,200-year-old model airplane had wings that angle slightly downward at the ends – exactly the same improved wing design that was first used in modern times on the supersonic Concorde!

Everyone agrees that humans have always been interested in flight. But now it appears that we creationists were right – we have always been smart enough to do something about our interest in flight as well!

Prayer: Dear Lord, help people to understand that they are not simply glorified animals but they are Your creations. Help us all to see that You made each one of us for a purpose and help me to live for Your purpose. In Jesus' Name. Amen.

A Diet of Oil

II Corinthians 12:9
"And he said unto me, My grace is sufficient for thee, for my strength is made perfect in weakness. Most gladly therefore will I rather glory in my infirmities, that the power of Christ may rest upon me."

While scientists often talk about how it takes millions of years to form oil and gas deposits, they say little about the fact that oil and natural gas are forming today in the ocean's sediments. The sediments of the Gulf of Mexico are well-known examples of the rapid natural formation of petroleum deposits. In addition, oil and natural gas are seeping from petroleum deposits under the floor of the Gulf of Mexico.

But even more amazing is the discovery that there are sea creatures that actually use crude oil and natural gas for food. Scientists have known that some microbes digest oil. But they have discovered new species of worms, clams, mussels and crabs that cluster around petroleum and natural gas seepages on the floor of the Gulf. So far they have found 30 sites off the coast of Louisiana and Texas where these strange creatures have set up communities around seepages.

While doing this deep-water research, scientists also discovered lakes under the ocean! They found patches of water much saltier and, therefore, much heavier than sea water resting on the sea floor like under-sea lakes. They report that the saltier water lakes actually have ripples on their surfaces, just like a lake on dry land.

Despite claims made by some scientists, we humans are still only in the early stages of finding everything there is to find in the creation. Discoveries like this can show us that we are a long, long way from even beginning to understand how things in the creation actually work.

Prayer: Father, You humble man because our out-of-control pride needs a better grasp of reality. Help me to see that without Your Son, my Lord and Savior Jesus Christ, I have nothing to boast about. Let my boasting be in Him. Amen.

REF: "Ocean depths harbor an ecosystem of creatures who dine on oil, natural gas." *Star Tribune* (Minneapolis), Dec. 31, 1989. p. 15A.

How Does the Nose Work?

Genesis 27:27
"And he came near, and kissed him: and he smelled the smell of his raiment, and blessed him, and said, See, the smell of my son is as the smell of a field which the LORD hath blessed."

Some things don't have to be understood to be appreciated. You can enjoy the smell of dinner cooking or the scent of a rose without any idea of how your nose works. And believe it or not, you're doing just as well as the most brilliant biologist.

While your nose knows how it works, science cannot explain just how we sense scents. It is known that inside our noses, behind the bridge of the nose, are cells that can sense smell. These cells are able to detect and identify airborne molecules from an open rose or a cooking roast. But no one knows just how these cells turn those molecules into the sense of smell that we experience.

To make matters more complicated, the sense of smell is one of our most complex senses. A single seemingly simple odor may contain more than 1,000 different chemicals. One sniff is likely to start activity all over the brain. Scientists have proven what experience has already shown most of us – a smell can also trigger emotions and memories, depending on an experience related to that smell. In addition, your sense of smell is linked to your sense of taste, which is why food can seem to be tasteless when you have a head cold.

A sense of smell has saved countless lives and brought joy and pleasure to all but those few whose sense of smell has malfunctioned. Yet it is so complex that modern science doesn't know how it works – yet another testimony to the wisdom of our loving Creator.

Prayer: Dear Father in heaven, I thank You for the sensation of smell and the help and pleasure it gives me. As everything You have given to me glorifies You in all things. In Jesus' Name. Amen.

Ref: Reyneri, Adriana. 1984. "The nose knows, but science doesn't." *Science 84*, September. p. 26.

The Big Universe

I Corinthians 15:41
"There is one glory of the sun, and another glory of the moon, and another glory of the stars: for one star differeth from another star in glory."

One of the most breathtaking sights any human being can experience is available for free. The night sky, especially away from city lights, is a sight that staggers the imagination.

It's no wonder humans are fascinated with the stars. This fascination explains why some of the earliest modern inventions were telescopes. It also explains why, even in the 17th century, the astronomer Kepler wrote to Galileo suggesting that humans might one day travel to the stars. And it explains why, within four centuries, some have actually set foot on the moon.

The night sky is filled with seemingly countless stars. Our space probes have given us dramatic pictures of the other planets. We have seen the largest volcano in the solar system on Mars, the dramatic volcanoes of Io, the wispy rings of the outer planets and the white, puffy clouds floating in the blue atmosphere of Neptune. Yet all of these planets and stars are visible to the naked eye. And what the unaided eye can see is less than 100-billionth of the universe! We can directly measure only distant objects that are 300 or so light years away – which is far less than we can see.

Close-up views of the objects in the night sky produce even more awe and wonder than the night sky itself. In a very real sense, the sky is the clearest and the most glorious of our Creator's fingerprints on His incredible workmanship!

Prayer: Dear Father in heaven, the heavens glorify You; the wonderful eyes with which I behold the heavens glorify You! Let me too glorify You among men, especially for Your greatest glory, Your plan of salvation through Jesus Christ! Amen.

Ref: Ferris, Timothy. 1984. "Spaceshots." *Science 84*, September. p. 60.

A Journey Through Inner Space

Matthew 6:25
"Therefore I say unto you, Take no thought for your life, what ye shall eat or what ye shall drink; nor yet for your body, what ye shall put on. Is not the life more than meat, and the body than raiment?"

What would the typical living cell look like if enlarged to the size of New York City? Let's take an imaginary journey through the cell.

As we approach the giant cell in our special ship, we would see breathtaking beauty, order and busyness within the cell. On the surface of the cell we see millions of openings stretching as far as the eye can see. Each one independently opens and closes, allowing materials to enter or leave the cell.

As we enter one of these vast portholes, we see an endless array of hallways leading in all directions. Some of these hallways lead to the cell's memory banks in the nucleus. There, all the activities we are about to see are not only regulated but also checked by very fussy quality controllers. Other halls lead to giant processing plants, while still others lead to assembly plants within the cell. If we move along a giant hallway to one of these assembly plants, we will see marvelous organization. Raw materials, like those we saw entering the cell, are sent here for assembly. Robot-like structures – really proteins – are working on the assembly line while other proteins, called enzymes, are supervising their work.

This view of the cell, provided by modern science, makes it even more difficult to believe that the cell, or any life, could have arisen from chance collisions of atoms. It is for this reason that several famous evolutionists have given up on the idea that chance and accident could have created life!

Prayer: Dear Lord Jesus, I marvel at how You, the Instrument Who created the living cell, became a living human being Yourself for my salvation. Though I cannot understand such love, help me never to take Your gift of salvation for granted. Amen.

Your Body's 100,000 Sentries

Psalms 139:14
"I will praise thee; for I am fearfully and wonderfully made: marvelous are thy works; and that my soul knoweth right well."

Despite the fact that most microorganisms are necessary and good, throughout our lifetimes each of us encounters tens of thousands of different infectious bacteria, viruses, fungi and parasites.

Even more remarkable is the fact that most of the time our immune systems disable these potentially lethal invaders before we ever show any symptoms of infection. At any given time more than 100,000 unique sentries posted throughout your body identify invaders, sound the alarm, and even issue specific chemical instructions for their destruction. These sentries may also be thought of as tiny doctors who identify a potential illness, discover the cure, and apply it even before the infection gets underway.

The immune system has puzzled scientists. Researchers know that our bodies do not keep a set of genetic blueprints for these sentries, which are called B cells. How then does our body make these sentries or develop the genetic information necessary to disable invaders? Researchers have learned that the body has a small library of DNA fragments that are continually being shuffled into new patterns so that the body is almost instantly ready for any invader.

The fact that even medical researchers are in awe over the design of our immune system verifies what the Bible says: "I am fearfully and wonderfully made."

Prayer: I thank You, dear Father, that I am fearfully and wonderfully made. Help me to take care of the temple of my body that You have given me so that I may be a fit and able instrument for You in this world. In Jesus' Name. Amen.

Ref: "MIT researchers isolate master-builder' disease-fighting gene." *Minneapolis Star-Tribune*, Dec. 22, 1989. p. 2.

A Monkey, a Bird and a Story

Romans 1:18-19
"For the wrath of God is revealed from heaven against all ungodliness and unrighteousness of men, who hold the truth in unrighteousness; because that which may be known of God is manifest in them, for God hath shewed it unto them."

Evolutionists say that human beings evolved from an ape-like creature. And how do they claim to know this? Because they find various kinds of apes in the fossil record. As one would expect, some of these apes are a little – and I stress, a little – more like us than others. They are all more like apes than like us. But evolutionists will tell us that those apes that are a little more like humans are evidence for evolution. Is this a logical way to think?

Some other cultures believe that human beings came from birds. Let's say that a monkey and a bird, both sitting in a tree, were fossilized along with the tree. If a paleontologist from a western culture dug up the fossilized tree, monkey and bird, and decided they were the right age, he would conclude that the monkey could have something to do with human evolution. He would reject the bird, concluding that it offered no proof at all for human evolution.

But what if the same fossilized tree, monkey and bird were discovered by someone who thought that humans came from birds? He would reject the monkey as proof of human evolution, possibly finding that the bird fit his theory very nicely. Notice that each one found the so-called "evidence" he was looking for. But what did they both actually find? What they really found were a fossilized tree, fossilized bird, and fossilized monkey. That's all. The rest was a story each made up to fit what he already believed before his discovery.

While science has made up lots of stories, it has never actually found anything that contradicts the Bible!

Prayer: Dear Father, man makes up stories so that he can claim that Your Word is nothing more than another made-up story. Help me to learn Your Word so well that I can easily see through all of the stories that man makes up. In Jesus' Name. Amen.

The Crafty Flea

I Peter 5:6-7
"Humble yourselves therefore under the mighty hand of God, that he may exalt you in due time: Casting all your care upon him; for he careth for you."

The water flea fills one of the lowest spots on the food chain. What this means is that even though the water flea is concerned about little else than going about his business, just about everyone else in the pond is interested in eating him.

Not much larger than a grain of rice, the water flea has been given such a clever defense against predators that even scientists were fooled for a while. Adult water fleas, or *Daphnia*, are egg-shaped except for their legs, and are easy to swallow. Newly hatched water fleas are usually egg-shaped, too – making them a tender, easy to swallow lunch. But as the number of predators in the pond or stream grows, hatchlings begin to develop an odd array of sharp projections on their bodies. Some look so different from their parents that at first scientists misidentified them as a different species.

The water fleas' strategy involves more than just trying to look like something different from a traditional water flea lunch. The sharp projections make them hard to swallow. Anyone would rather swallow an egg than a porcupine! Research shows that these projections reduce mortality of young water fleas by 50%.

If the Creator takes this personalized care of water fleas, how much more He wishes to take this kind of personalized care of you! None of us need ever feel alone or uncared for. Your Creator seeks a personal relationship with you through His Son Jesus Christ!

Prayer: *Dear Father in heaven, I thank You that You have not left me alone and without hope. Renew and strengthen me in a closer relationship to You by means of Your loving invitation in Your Word – the Bible. For Jesus' sake. Amen.*

REF: Fellman, Bruce. "Quick-change flea." *Science 84*. p. 76.

Electric Skin

Job 42:2-3
"I know that thou canst do every thing; and that no thought can be withholden from thee. Who is he that hideth counsel without knowledge? therefore have I uttered that I understood not; things too wonderful for me, which I knew not."

It's weatherproof, as flexible as paper, can generate electricity, serve as a microphone or radio speaker, sense your presence, act like a sponge, and has some abilities that are like human skin. What could this strange material be? It's a plastic called PVDF.

PVDF has so many amazing abilities that one scientist remarked, "God was good to us." Those automobile stickers saying "unleaded fuel only" are made of PVDF. It has also been used to coat skyscraper surfaces to protect them from the weather. If you attach a few wires to the film, it will serve as a microphone. Or if you attach a speaker wire to it, it will serve as a speaker. It can sense weight, movement and infrared radiation. When bent, it generates electricity. If 220 tons of the material was formed into sheets that can be moved by sea waves, it would generate one megawatt of electricity.

It is this ability to generate electricity that makes it like your skin. Human skin and PVDF both act like an electric sponge. When you push on the sponge, electricity flows out just like water. When you release pressure on the sponge, electricity flows back in, just as water is soaked up by a sponge. Not only does this help us feel things on our skin, but PVDF can be used in the same way to give machines the ability to feel things they touch.

The abilities of PVDF are amazing scientists. Yet its abilities are much less significant than the abilities of the skin that God has given us.

Prayer: Dear Lord, I am surrounded by so many of your blessings that I cannot think of them all, no matter how long I try. For this I thank You and ask that I would continue to grow in my ability to see Your wonderful handiwork. In Jesus' Name. Amen.

Ref: Amato, Ivan. 1989. "Fantastic plastic." *Science News*, v. 136, Nov. 18. p. 328.

Population and the Age of the Earth

Genesis 6:8
"But Noah found grace in the eyes of the LORD."

How long have people been living on Earth? The evolutionist says millions of years. Bible-believing Christians generally say only about 6,000 years. But the answer to this question is amazingly simple.

If we start with only two people, and they have four children who live to have their own children, the second generation now has twice as many people – four. Now allowing for infant mortality and other human problems that keep population down, we still find that on the average it only takes about 130 years to double the Earth's population. This figure fits into known historical records. And if anything, it's a conservative number.

If human history is 2 million years, as the evolutionists say, the Earth ought to have a lot more people than it does now. Alternatively, if we accept the 2 million years, then it must have taken 125,000 years to double the population in order to finish with today's world population. But that doesn't make any sense at all, especially since human historical records show that the doubling time is about 1,000 times less!

But if we start with eight people and reckon that the population doubles every 130 years, we find that it takes only about 4,000 to 4,500 years to get a population of 1 billion. And that was the Earth's population in the year 1800 – just about 4,200 years after the Flood – through which only eight people were saved to repopulate the Earth!

Prayer: Dear Lord, even the growth of human population testifies to the truth of Your Word! Help me to remember that to You humanity is not a mass of people, but that even though it numbers in the billions, each is an individual whom You are seeking with Your Word. For Jesus' sake. Amen.

Ref: Weigand, Cleone H. 1985. "Morality remains the best way to stem population growth." *Milwaukee Journal*, Apr. 14.

A 4,000-Year-Old Computer Language

John 1:1
"In the beginning was the Word, and the Word was with God, and the Word was God."

The language is so logical and precisely structured that it is being used to create programming language that a computer can understand. Unlike most languages of the world, this language has few irregular verbs, no prepositions and almost no exceptions to strict grammatical rules. What would take a long sentence to say in most languages can be said in only a few words in this language.

One would think that we are talking about a new language, carefully designed and crafted by modern scientists. But this language, called Aymara, has been spoken by peasants in the Andes for 4,000 years! Because of its highly logical structure, Aymara is being used as a bridge language that enables computers to translate one language to another.

Since 1986 a computer has been using Aymara to translate Spanish technical and legal documents into English in the office of the Panama Canal Commission. This amazing computer can translate a 60,000-word document with 80% accuracy in only an hour. The final product must then be reviewed by a translator.

Generally speaking, modern languages are less logically structured and more irregular and often cannot express as many shades of meaning as can ancient languages. In other words, as we go back in time, languages are not less structured and less expressive, but more highly structured and more expressive than today's languages. This is powerful evidence that language has not come from grunting cavemen, but from the Creator Himself!

Prayer: Dear Lord, the Word through Whom all things were created, I thank You for the gift of language. Help me to be faithful to You in my use of language, not only in speaking the truth in love and avoiding falsehoods, but also in telling others what You have done for them in Your death and Resurrection. Amen.

Ref: "Computer application helps save an ancient language." *Minneapolis Star Tribune*, Feb. 24, 1990. p. 9A.

Does the Lion's Tooth Bite Your Lawn?

Roman 8:28
"And we know that all things work together for good to them that love God, to them who are the called according to his purpose."

As spring comes to the temperate areas of the northern hemisphere, one of the first signs of spring is the blossom of the Lion's Tooth. This bright yellow flower is also call the Irish daisy, priest's crown, peasant's cloak and yellow gowan. Its other name appears on the sides of bags of lawn care chemicals – the dandelion.

The millions of homeowners who wage unrelenting war on the dandelion will probably not be pleased to hear anything good said about the plant. But the dandelion actually has a very interesting history. It was first brought to North America from Europe because of its beautiful flower. As late as the early 20th century seed companies were still selling packets of dandelion seeds for home flower gardens.

Up to 1957, more than 100,000 pounds of dandelion roots were imported annually to the United States for pharmaceutical use. In spring the dandelion contains mannitol, which is used as a base for pills, a treatment for hypertension and coronary insufficiency, as well as in manufacturing radio condensers and percussion caps. And many people enjoy a salad that includes young dandelion leaves freshly picked in the spring.

While you may not like dandelions in your lawn, the dandelion illustrates that God has not made anything that is without use. It is up to us to put what He has given each of us to use for His purposes.

> **Prayer: Dear Lord, forgive me for the times that I have not appreciated the good purpose for which You have brought some experience or event into my life. Through Jesus Christ, fill me with the conviction that my life can fulfill Your purpose if I follow You. In His Name. Amen.**

Ref: "A weed by any other name." *Science 83*, Apr. 1983. p. 82.

The Creator Rebuilds Lives

Hosea 6:6
"For I desired mercy, and not sacrifice; and the knowledge of God more than burnt offerings."

For generations we have heard that people become criminals and go to prison because of their parents, friends, poverty or society itself. Yet there have always been a majority of people who come from bad homes or poor neighborhoods who have lived responsible lives.

Now some police officials and psychologists are suggesting that criminal behavior is the fault of the criminal himself. The lifelong criminal chooses a pattern of dealing with life that is different from most people – and it's often evident by the time he is four years old. They characterize what they call the criminal mind as a person who chooses to lie instead of taking responsibility. They say that traditional explanations that blame society for crime simply help the criminal mind avoid responsibility. The career criminal likes to break the law. As one rapist said, "If rape were legalized, I'd do something else."

No, the origins issue is not simply a philosophical debate! We have heard from many serving prison terms for serious crimes. Those who contact us all say the same thing. They tell us that they have realized they must become responsible for their lives. They realize that in order to learn a new way of thinking, they must begin by learning how to approach the Creator Who made them. And we are pleased to help them learn more about Him and His inviting and all-forgiving love to them in Jesus Christ.

> **Prayer: Dear Father, because I, too, sin, I know that the temptation to avoid responsibility for my actions in strong. Through Your Word of promise in the gospel, remind me that You are always more pleased to forgive me for Jesus' sake than have me try to avoid You. Let my peace be in Christ. Amen.**

Ref: "Exposing the criminal mind." *Science 84*, Sept. 1984. p. 84.

The Ultimate Engineer

Job 38:4, 36
"Where wast thou when I laid the foundations of the earth? declare if thou hast understanding. Who hath put wisdom in the inward parts? Or who hath given understanding to the heart?"

Did you ever wonder why, while there are lots of round and cylindrical living things, there are almost no square plants or animals? Why isn't there any animal that has a skeleton made out of metal? And while there are so many ways for living things to move about, why do almost none of them have wheels?

We make lots of square things, use metal frames in the things we build, and use wheels on lots of moving things. But these features don't offer good solutions to the problems most living things have to deal with in life. Wheels are useless for going through the jungle, climbing trees, flying or burrowing. In engineering language, all living things show a high degree of design sophistication.

For example, the skeletons of all mammals have a ratio of 30% shock-absorbing collagen to 70% calcium phosphate for strength. This ratio provides the very best balance for holding up a mammal's weight during locomotion. Engineers also know that in order to get the best flow of a liquid – such as blood – a pipe's radius squared is to be equal to the sum of the radius' squared of the branches. And this is exactly the relationship found in all living and fossilized creatures, from sponges to humans!

The impressive engineering found in all living things – and even in the oldest fossils – offers elegant testimony to the Creator's wisdom and power! None of us should be shy about recognizing Him when we are with others.

Prayer: Dear Father in heaven, I thank You that You have not left the world without witness to Your great wisdom and power. Make me a better witness to You and what You have done for me through Your Son, my Savior, Jesus Christ. In His Name. Amen.

Ref: Wickelgren, Ingrid. 1989. "The mechanics of natural success." *Science News,* v. 135, June 17. p. 376.

Where Is the Garden of Eden?

Genesis 2:8
"And the LORD God planted a garden eastward in Eden, and there he put the man whom he had formed."

One mystery that has intrigued humans for thousands of years is the location of the Garden of Eden. Even Sunday school children stare at maps and carefully read Genesis 2:8-14, trying to locate it. But when we wonder where the Garden of Eden was, there are other passages in the Bible that we need to consider. Genesis 3:23-24 makes it very clear that God closed the Garden to humans – meaning no one could ever go back.

But there is even more important information in Scripture on the location of the Garden. Genesis 7:11 says that "all the fountains of the great deep were broken up, and the windows of heaven were opened." These words about the breaking up of all the fountains of the deep suggest violent and widespread volcanic action. The geological record says that such violent action was indeed widespread.

And very clearly the opening of the windows of heaven describes much more than just heavy rain. In the Genesis Flood, the entire Earth's surface underwent violent reshaping. We find evidence of this in fossils that are found today on mountaintops. We know that some of these creatures lived five miles below the ocean's surface!

You can find rivers on maps today with the same names as those mentioned in Genesis. But the fact is, those rivers on our maps today are only named after their pre-Flood counterparts. God closed Eden to sinful humanity and violently reshaped the Earth's surface in the Flood. The only way back to God is through Jesus Christ.

Prayer: Dear Lord, help me to abandon all hope of finding You by my own efforts and instead trust only in what Jesus Christ has already completed for me on Calvary. Help me to teach others to do the same. In His Name. Amen.

Inviting Ants

Proverbs 6:6-8
"Go to the ant, thou sluggard; consider her ways, and be wise: Which having no guide, overseer, or ruler, Provideth her meat in the summer, and gathereth her food in the harvest."

Most of us know that many plants depend on birds, animals and the wind to scatter their seeds. This is important, because most seedlings do better when they have a chance to sprout and grow away from the parent plant and other seedlings.

But some plants work very hard to invite ants to scatter their seeds. One example is the bloodroot. When the mature bloodroot berry splits, it reveals a seed that is wrapped in what looks like a white tissue. This "tissue" is called an *elaiosome*. The elaiosome contains food materials that the ants need but cannot make for themselves. The ants take the seed, still wrapped in the elaiosome, back to their nest where they remove the elaiosome. The ants don't eat the seed but, instead, deposit it on the nest's moist compost heap, where the young plant gets a good start when the seed sprouts.

While the trillium relies on birds to scatter the seeds in its red berries, it also counts on ants to scatter any seeds the birds leave behind. When the berry walls decay, seeds are released wrapped in ant-attracting elaiosome. Some kinds of ants in Australia remove the elaiosome from seeds and then bury the seeds just deep enough to protect them from dangerous temperatures, but shallow enough to receive the heat they need for germination!

This important relationship between certain plants and ants, and the ants' knowledge of farming, speaks clearly *for* the intelligence and wisdom of the Creator over all of the creation and *against* any explanation that says this just happened all by itself.

Prayer: Dear Lord, You have filled the creation with relationships between different living things. Help me always to place the highest value of all on my relationship with You through Your Son Jesus Christ. Amen.

Ref: Beattie, Andrew J. 1990. "Ant plantation." *Natural History*, Feb. p. 10.

Science and Miracles

Luke 1:37
"For with God nothing shall be impossible."

Do you realize that the Bible's own arguments have been stolen by evolutionary scientists in their attempt to prove that there are no such things as miracles? And it's past time that Christians reclaim those arguments in order to deflate the doubt that has been created.

Evolutionists say that since the world is an orderly place with predictable physical laws, we know that miracles can't happen. And they're right on everything but their conclusion. It is because the world is an orderly place with predictable physical laws that miracles are important.

The Bible clearly teaches that the world follows orderly laws. For example, Genesis 1:11 tells us that plants reproduce in an orderly predictable pattern – after their kind. In Genesis 1:14 we read about how the stars move across the sky in an orderly pattern so that we can tell what season it is. And if the world were not an orderly, predictable place, we couldn't tell if something was a miracle or not. In a world where things happen willy-nilly, miracles would have no meaning.

It is only in a predictable world that science can study where miracles can be what the New Testament calls them – signs. So the very reason evolutionists use to deny the existence of miracles is really the only reason we have to accept the possibility – and the reality – of the Creator's miraculous working in His creation!

> ***Prayer:*** *Father, ours is an age of doubt. Forgive me for the times I have been intimidated by the arguments of unbelieving scientists. Help me to see through their poor arguments and uphold Your truth before men. In Jesus' Name. Amen.*

Ref: Bartz, Paul A. 1990. "The Bible teaches a predictable world — but not naturalism." *Bible-Science Newsletter*, Mar. p. 10.

Darwin's Child Murdered!

Psalm 40:5
"Many, O LORD my God, are thy wonderful works which thou hast done, and thy thoughts which are to us-ward: they cannot be reckoned up in order unto thee: if I would declare and speak of them, they are more than can be numbered."

Evolution is often challenged with examples that show that the creation is carefully and lovingly designed by the Creator. Many of these evidences leave evolutionists silent and unable to respond. But one evidence for design is so powerful, it almost seems unfair for creationists to mention it. Yet they do, as in one release from scientists at the Institute for Creation Research.

While the brain weighs only three pounds, it can do the work of 1,000 super-computers. It doesn't need to be connected to a power source and it doesn't overheat because it is able to make its own electricity and it operates on only microvolts of power. If your brain's 10 trillion cells were placed end-to-end, they would stretch for over 100,000 miles. Your brain has the capacity to store every word of every book on a bookshelf 500 miles long.

In order for the human brain to have evolved from a simpler brain in the time that evolutionists claim it has, the brain would have had to evolve millions of new cells every year for millions of years. A.R. Wallace, co-discoverer with Charles Darwin of natural selection, once noted that there is a huge gulf between the human brain and that of the ape. Darwin recognized what Wallace's argument did to their theory and responded, "I hope you have not murdered completely your own and my child."

God has given you a brain that has wonderful abilities. Don't waste one of the greatest material gifts the Creator has given you.

Prayer: Forgive me, Father, for underestimating the great gift that You have given me – my brain. Grant me Your Holy Spirit so that I may begin and continue all learning and wisdom in You. In Jesus' Name. Amen.

Ref: DeYoung, Don, and Richard Bliss. 1990. "Thinking about the brain." *ICR Impact*. Feb. p. 1.

The Destructive Power of Water

II Peter 2:4-5
"For if God spared not the angels that sinned, but cast them down to hell, and delivered them into chains of darkness, to be reserved unto judgment; And spared not the old world, but saved Noah the eighth person, a preacher of righteousness, bringing in the flood upon the world of the ungodly;"

Many of the features of the Earth's surface have been formed by the cutting and eroding action of moving water. When you think about how hard rock is compared to water, it's easy to believe that it must have taken hundreds of thousands or even millions of years for water to shape the land. This ignorance about how rapidly water cuts rock has cost people their lives.

In April 1987 a 300-foot section of the 540-foot-long New York Thruway bridge collapsed into the Choharie Creek. The moving waters of the creek had created turbulence around the bridge's pilings. Within a few years this turbulence cut away the rock in which the pilings were anchored, and with nothing to hold it up, the bridge collapsed.

In June 1987 a section of the 2,800-foot-long Clearwater Pass Bridge in Florida dropped 10 inches. Divers sent down to inspect the bridge pilings found that more than 10 feet of rock had been scoured away from the bridge's pilings. Similar instances of moving water cutting away solid rock in a short period of time can be found around dams.

When many of these structures were built, it was assumed that it took tens of thousands of years for water to erode solid rock. But experience has now shown us that water is able to do in a few years or even a few hours what scientists once thought took thousands or millions of years. Another lesson to be learned is that we don't need to claim that it took tens of thousands of years to form the water-carved features of our Earth!

Prayer: Dear Heavenly Father, help more people to see that Your Word is truth and that it has not been shown wrong by man's science. Make it increasingly difficult for people to ignore the Bible, which tells us of Your plan of salvation for us through Jesus Christ. Amen.

Science Sheds Light on the Darkest Day

Matthew 27:45
"Now from the sixth hour there was darkness over all the land unto the ninth hour."

Historical science is one branch of human knowledge that consistently supports the biblical record.

In Matthew 27:45 we read that darkness covered the land from 12:00 noon until 3:00 in the afternoon, while Christ hung dying on the cross for our sins. The biblical language that describes this darkness makes it clear that it was not due to clouds. Nor could this darkness be caused by an eclipse of the sun. Christ's crucifixion took place at Passover, a time when the moon is full. Besides, a solar eclipse never lasts for three hours.

It is interesting to note that this darkness was also recorded in Egypt. According to one ancient record, as the same darkness descended upon Egypt, it was so fearfully dark that Dinoysius of Egypt exclaimed, "Either the God of Nature is suffering, or the machine of the world is tumbling into ruin."

Indeed the God of nature, the Word who created everything that exists, had left His heavenly glory to repair the damage humanity had done to His work. He took our form upon Himself so that He could carry our sins. And as He experienced the full pain of God's just punishment for our sin in His perfect body and perfect mind, the God of nature was truly suffering and even dying. His burden was our sin, so His death gives us life and His resurrection makes that life worthwhile because it is, once again, a daily walk with our Creator. Is there darkness and hopelessness in your life? Come to His cross and know that there He has restored you to your Creator.

> ***Prayer: Lord Jesus Christ, I thank You that You left the glories of heaven to carry my sin on the cross, offering Your perfect life for my sinful life so that I may no longer be separated from You. Let me never treat Your love lightly. Amen.***

Ref: Corliss, William R. 1983. "Tornados, Dark Days, Anomalous Precipitation, and Related Weather Phenomena." Glen Arm, MD: *The Sourcebook Project*. p. 31.

The Miraculous Seed

I Corinthians 15:35-38
"But some man will say, How are the dead raised up? and with what body do they come? Thou fool, that which thou sowest is not quickened, except it die: And that which thou sowest, thou sowest not that body that shall be, but bare grain, it may chance of wheat, or of some other grain:"

Can the dead come back to life? While many scientists assure us that they cannot, the empty grave of Christ assures us that in Him everyone who trusts in His saving work on the cross of Calvary will live bodily with Him forever. When modern science tries to raise doubts about the bodily resurrection of Christ, just keep in mind that science cannot even explain the miracle of the seed.

Any seed, whether an ordinary bean seed, an acorn or a coconut, is a marvel of life. Consider the oak tree. Within the acorn, weighing less than an ounce, is a tiny embryonic oak tree, enough nutrition to get the young tree started on its own and all of the coded information needed not only to direct the start of growth, but to also guide the design and development of tons and tons of tree! Lotus seeds can remain alive and sprout for up to 1,000 years!

If it had been up to the first plant to figure out how to accomplish this miracle by trial and error, we would still be waiting for the first seed to develop! But Scripture solves this problem for us by clearly stating that God is the author of all life! Besides Scripture's clear statement on the matter, it seems most unscientific to say that the tiny, packaged bundle of life we call a seed was not carefully designed and created but came about by chance.

And as the seed in the ground shakes off its old, dry coat to burst forth in vibrant new life, let each of us look to Christ to shake off the old and decayed and allow Him to place on us the vibrant new life He gives us through His resurrection.

Prayer: Dear Lord Jesus Christ, Your empty grave speaks louder than all the words of doubters. Let me truly show forth the new life You have given me so that the hearts of doubters may be opened to You. Amen.

Ref: Pope, Jon Cedar. 1979. "Carrying an immature plant through time and space." *Science Digest*, May. p. 60.

"Just So" Stories of Evolution

I Timothy 4:16
"Take heed unto thyself, and unto the doctrine; continue in them: for in doing this thou shalt both save thyself, and them that hear thee."

There is a common sense saying in science that when anything can serve as an explanation to support a theory, the theory is not scientific.

Let's say that you have a friend who has a theory that invisible pink elephants fly over his house every day at noon. He wants to prove his theory to you. So you agree to be at his house at noon to see his evidence. It's noon and you are staring at the sky, seeing nothing but a nice, clear blue sky – the very sight you have expected and hoped to see. At one minute past noon he asks you, "Well, did you see them?" You answer "no," and he responds, "That proves it. I told you they were invisible."

When science discovered that the human brain is divided into specialized left and right halves, evolutionists said this specialization shows why humans have evolved further than the animals and can do so many things. Evolutionists said the left side of the brain became specialized so that humans could specialize in speech while also specializing in emotion with the right side of the brain. Then scientists discovered that many animals also have this left and right brain specialization. All of a sudden, the fact of left and right brain specialization among so many creatures was used by evolutionists as evidence of our supposed relationship with animals.

Evolution is a "just so" story where anything can serve as "proof." But when anything can be proof, then nothing is proven.

> ***Prayer: Dear Lord, people use many stories to convince themselves that they can ignore Your true Word. Help me to see that I too often use the story that "I just don't have time to read the Bible today," repent of it and resolve to hear You teach me every day in Your Word. Amen.***

Infants' Talk Puzzles Scientists

Luke 10:16
"He that heareth you heareth me; and he that despiseth you despiseth me; and he that despiseth me despiseth him that sent me."

    Because we were created by God who values communication and has always desired to communicate with us, we are born with the natural ability to learn and to speak language. Those who believe in evolution usually disbelieve that humans have the natural mental ability to speak, since they think both the ability to speak and language itself evolved.

    New research has forced evolutionists to change their thinking so that it is more in line with what the Bible says. As one science writer put it, learning to speak is "a miracle." Instead of being a so-called "blank slate," infants are born into the world with the inborn mental and physical abilities to learn any language on Earth. Researchers have found that the ability to learn language is highest before a child reaches three years of age. Researchers learned that by age three, children from bilingual households have learned the basics of, and the differences between, the two languages. A child's brain is – to use computer language – specifically wired to learn language.

    Infants understand language long before they have mastered the breath, mouth and tongue control to speak it. All those baby noises are preprogrammed exercises needed to master control of our speaking abilities.

    The fact that our Creator built in our ability to communicate helps to show us that we were intended to have a communicating relationship with Him. He speaks to us in the Bible and we speak to Him in prayer. Do you have good communication with Your Creator?

Prayer: Dear Heavenly Father, I confess that communications between us have not been as good as they should be. It is all my fault. Fill me with eagerness to hear Your voice in the Bible and to come to You in prayer. In Jesus' Name. Amen.

Ref: Kobren, Gerri. 1990. "Tots assemble building blocks of language." *Minneapolis Star Tribune*, Jan. 20. p. 1E.

U.S. Army Celebrates Its 75,000 Birthday!

Romans 1:22
"Professing themselves to be wise, they became fools..."

We are often asked, where do evolutionists get all those millions of years? It seems they are always finding rocks and fossils that are tens or hundreds of thousands or even millions of years old.

Some years ago scientists discovered human bones in California buried under thick layers of mud. Scientific authorities studied the bones and dated them at about 75,000 years old. Digging deeper, scientists discovered an old United States Army button! Was the U.S. Army around 75,000 years ago, or was the dating method unreliable? More recently an archaeological team from a Japanese university discovered drawings on a cave wall on a Japanese island. This important discovery was dated at 10,000 to 13,000 years old. When one of the local residents of the island heard about the discovery, he stepped forward to confess that as a boy he often drew on the walls of the cave with charcoal.

Can you imagine the misery we would have if modern medicine was only as reliable as these dating methods?

The simple answer to where evolutionary scientists get all those years is that they make them up. There is no machine or scientific magic that tells a scientist that a rock or fossil is tens of thousands or millions of years old. The oldest and most accurate record of ancient times that we humans have is the Bible, and the Bible leaves no room for tens of thousands or millions of years!

> *Prayer: Dear Lord, call back those millions who have been falsely led to believe that the Bible is not trustworthy because modern science has disproven the Bible as Your historically accurate record. Help them to see that evolution is simply a false religion that would replace You, the Creator, with the creation itself. In Jesus' Name. Amen.*

Ref: Jackson, Wayne. 1990. "Scientific red faces." *Reasoning from Revelation,* v. II, n. 1, Jan. p. 3.

Seeing Colors

II Corinthians 4:3-4
"But if our gospel be hid, it is hid to them that are lost: In whom the god of this world hath blinded the minds of them which believe not, lest the light of the glorious gospel of Christ, who is the image of God, should shine unto them."

The eye, with its incredibly sensitive color vision, is increasingly being recognized as a marvel as we learn more about how it works. The eye is able to detect the smallest measure of light known to physicists. These units are called photons.

The normal human eye detects color using principles similar to those used by our color television. Within the eye are three kinds of light-absorbing molecules. They are found in the millions of cone-shaped cells in the retina. Each one of these three molecules absorbs one of the primary colors – red, green or blue. How could "natural forces" have accidentally stumbled onto the very principle of physics upon which color rests?

The sensitivity required for color vision is also astounding. There is only 75-millionths of a meter difference between the wave lengths of blue and green light. This size difference is so small that it could not be noticed even under the most powerful light microscope. Yet the eye has no trouble at all detecting the difference.

In his book *On the Origin of Species*, Darwin wrote, "To suppose that the eye, with all its inimitable contrivances for adjusting focus to different distances, for admitting different amounts of light and for the correction of spherical and chromatic aberration, could have been formed by natural selection, seems, I freely confess, absurd in the highest degree." On this one point we agree most certainly with Charles Darwin.

Prayer: I thank You, dear Lord, for the ability to see the beauty and wonder of Your creation. Give sight to those who are spiritually blinded by the idea that impersonal evolution could have designed our ability to see so that seeing You, they may be led to a relationship with You through our Savior, Jesus Christ. In His Name. Amen.

A Confused Flower?

I Chronicles 16:9
"Sing unto him, sing psalms unto him; talk ye of all his wondrous works."

We regularly look at unusual plants and animals that show the unlimited inventiveness of our Creator. Unusual living things are also fascinating because, since they are clearly unrelated to any other living things, each one is a challenge to evolutionary theory, which says that every living thing is related to other living things.

Botanists recently identified a small flower in Mexico that has left them stumped. Flowers have a pistil or female flower part made up of a stigma, style and ovary, where the seeds develop. This arrangement is surrounded by male pollen-producing stamens.

But the small flower they recently identified in southern Mexico has only one stamen, which is surrounded by about 50 pistils. It is so unlike any other plant that it took botanists a year to identify. Not only is the flower so unlike any other flower that it is obviously not related to any known flower, but botanists also admit that they are at a complete loss to explain the flower in evolutionary terms.

However, we don't have any problems explaining the flower, because we know from the Bible that there is no limit to God's imagination and power. And I suspect that He placed so many of these very different creatures in the creation to show modern humans that life has not evolved, but was created by Him. His hope is that, seeing Him, people might want to find out about a relationship with Him through His Son, Jesus Christ.

Prayer: I thank You, Father, for Your desire that I come to know You through Your Son, Jesus Christ. I ask that because I know You and what You have done for me, You would show me opportunities to tell others about Your inviting grace. Amen.

Ref: "Very small flower found in Mexican jungle has its sex orientation reversed." *Minneapolis Star Tribune*, Feb. 4, 1990. p. 5E.

Could Creationism Correct Science?

Job 42:1-2

"Then Job answered the LORD, and said, I know that thou canst do every thing, and that no thought can be withholden from thee."

In the secular media we often hear spokespersons for evolution say that creationism would ruin science. But several years ago an evolutionist suggested in his paper that creationism may actually help correct science.

According to the article, astronomer Alan Batten said in a 1984 paper in the *Journal of the Royal Astronomical Society of Canada*, "My impression is that many of the biologists who write about evolution for the ordinary public are committed in their minds to the belief that the question 'Why?' has no answer…[They believe] evolution is purposeless, and life and Man have arisen only by chance… this is a legitimate belief for a person to hold, but that it is required by the scientific evidence is highly questionable. To hold it as such is as much an act of faith as to maintain the opposite."

Batten encourages evolutionists to stop arguing scientific evidence with creationists, saying, "I believe we shall achieve much more by frankly admitting the limitations of scientific research and concede that even creationists are raising legitimate questions."

Those scientists who are creationists object to dogmatic claims that "evolution is a fact." Batten is simply asking for honesty about evolution among his fellow scientists. If his suggestions were taken literally, both education and science would benefit and be lot less offensive to Bible-believing Christians.

Prayer: Dear Lord, as Creator it must sadden You deeply when You see how man seeks to hide from You through stories about evolution and how those who make up these stories are so unfair to Your people. Replace our discouragement with boldness to continue witnessing to Your truth. Amen.

Ref: Stahl, Philip A. 1990. "The God factor." *Astronomy*, Mar. p. 86.

Magnetic Birds?

Matthew 10:29
"Are not two sparrows sold for a farthing? and one of them shall not fall on the ground without your Father."

Most of us have heard a few of the theories that are offered to explain how birds are able to migrate for thousands of miles to an exact spot. The arctic tern migrates 22,000 miles a year to winter in the same spot where it wintered the previous year. But most birds seem to have this amazing ability even if they don't use it as often or to travel as far.

In one study scientists took a sea bird called the Manx shearwater from its nest on the coast of Wales to Boston and released it. The route back home across the featureless Atlantic is not a familiar one to the shearwater. Yet twelve and one-half days later the shearwater showed up back at its nest, over 3,000 miles from Boston.

The most dramatic research in recent years led to the discovery that many birds have a small amount of magnetic material in their brains, which seems to act like a built-in compass to help them tell where they are and where they are going. But additional research has shown that birds usually use more than just one method to navigate. They also use the sun, stars, changes in barometric pressure, low frequency sounds made by the wind and sea, and even odors.

The birds' ability to navigate using only one of these methods would be amazing enough. But the fact that birds have several methods available to them is a testimony to their Creator, who provided them with back-up methods as well. If no detail of a small bird's need is too unimportant for His attention, imagine how much more He is concerned about the details of our lives.

Prayer: Dear Father, forgive me for those times when I have thought that some detail that troubled me was too small to bring to You. I know, because You have told me so in the Bible that You desire an even closer relationship with me than we now enjoy. Lord, I believe. Help my unbelief. In Jesus' Name. Amen.

Ref: Cook, Patrick. "How do birds find where they're going?" *Science* 84. p. 26.

Is the Shark Related to the Pig?

Genesis 1:21
"And God created great whales, and every living creature that moveth, which the waters brought forth abundantly, after their kind, and every winged fowl after his kind: and God saw that it was good."

If all life evolved from, and is related to, earlier forms of life, then closely related animals should also have similar kinds of chemicals in their bodies. As the science of biochemistry has grown, those who believe in evolution have held high hopes that creatures that look similar to each other would also have similar chemicals. For example, insulin from a shark should be more like insulin from other fish than insulin from mammals.

Unfortunately for the evolutionist, chemical comparisons between various creatures don't usually show the same relationships as their evolutionary charts. For example, research at the Medical University of South Carolina compared the hormone relaxin that was produced by pigs and rats with relaxin produced by sharks. The results showed that the pig's relaxin was more like the shark's relaxin than it was like a rat's.

This seems to suggest that pigs are more closely related to sharks than to another mammal, the rat. A similar study comparing insulin showed a closer similarity between the shark and the pig than between the shark and another fish, the carp.

While evolutionists will sometimes point out instances where these comparisons have worked out as they expected, those results are unusual. Normally these comparisons show that the evolutionary histories and relationships that evolutionists claim are nothing more than imagination. As the Bible says, each kind of creature gives evidence to the fact that it was uniquely created by God.

Prayer: Dear Heavenly Father, I thank You that You have provided each living thing, including myself, with the unique bodily needs so necessary for life. I ask that the fact that each kind of creature has been specially created by You would become increasingly apparent to modern science. In Jesus' Name. Amen.

Plant Mathematicians

Romans 1:20
"For the invisible things of him from the creation of the world are clearly seen, being understood by the things that are made, even his eternal power and Godhead; so that they are without excuse:"

Throughout the centuries people have noticed, whether or not they believed in a Creator, that there is a mathematically precise structure to the universe and everything in it.

One everyday example of this precision can be found in plants. Many plants, including elm trees, grow their leaves, twigs and branches placed exactly halfway around the stem from each other. Next in the series are plants like the beech tree, with leaves placed one-third of the way around the stem from the previous leaves. Third in the series are plants like the oak, with leaves placed at two-fifths of a turn. The holly plant is next at three-eighths, then larches at five-thirteenths – and the sequence goes on.

Notice the number sequence of these fractions: 1,1,2,3,5,8,13, and so on. Each number is the sum of the two numbers that come just before it in the sequence. This particular mathematical pattern is called the Fibonacci series and is recognized as a basic mathematical series. Such mathematical precision is not arrived at by accident.

Such mathematical precision is only the product of power and intelligence, even as Paul says in Romans 1: "What may be known about God is manifest ... His invisible attributes are clearly seen, being understood by the things that are made." And while they are not flocking to the Bible, this is why many scientists are abandoning evolution!

Prayer: Dear Father, You are more than Creator. I thank You that You so loved the world that You sent Your only Son that whoever should believe in Him will have eternal life. Use the voice of Your people, beginning with me, to make the witness of Your love to man complete. In Jesus' Name. Amen.

Ref: Murchie, Guy. 1979. "The exquisite mathematics of nature..." *Science Digest*, Apr. p. 48.

The Universe's Missing Link

Hebrews 3:4
"For every house is builded by some man; but he that built all things is God."

Scientists reported in November of 1989 that a satellite that was supposed to help them answer questions about how the universe got here gave them no answers at all. The satellite, called the Cosmic Background Explorer, worked fine. Scientists just didn't find the evidences they expected to support their explanation of how the Big Bang could have produced the present universe.

The Cosmic Background Explorer failed to detect the patterns of microwave and infrared radiation that would support the Big Bang. "I am completely mystified how the present structures [in the universe] came to exist without leaving some trace," said John Mather, an astronomer with NASA's Goddard Space Flight Center.

Scientists noted that their failure to find the expected patterns of radiation leads to the obvious conclusion that neither we nor the universe exists. But Mather added, "We have not eliminated our existence yet."

Christians who know what the Creator tells us about the origin of the universe can easily see the problem here. The expected evidence for how the universe got here through a Big Bang isn't to be found, since the creation wasn't formed by a big explosion in space. The Bible says very clearly that this world didn't come into existence without a creator. Hebrews 3:4 tells us: "For every house is built by someone, but the builder of all things is God." It's not with satellites that we find God, but in His Word – the Bible.

> ***Prayer: Dear Heavenly Father, –we who know You often look for You every where but where we can find You revealed to us in more detail than we can understand – Your Word. Remind me to make daily use of Your Word in order to learn more of the relationship You would have with me through Your Son, Jesus Christ. In His Name. Amen.***

Ref: "Scientists still in dark on cosmos." *The Florida Times-Union*, Jan. 14, 1990. p. A-7.

Well-Designed Snails

Colossians 1:9-10
"For this cause we also, since the day we heard it, do not cease to pray for you, and to desire that ye might be filled with the knowledge of his will in all wisdom and spiritual understanding; That ye might walk worthy of the Lord, unto all pleasing, being fruitful in every good work, and increasing in the knowledge of God;"

Each one of us can learn about God's way of working with us from a fresh water snail that is often preyed upon by crayfish.

When life is easy and there are no crayfish around to threaten the snail, it usually lives only three to five months and grows to only one-sixteenth of an inch. But if crayfish are around to threaten and eat the snails, they will grow two to three times as large and live two to three times as long. This not only helps ensure that the snail population continues, but it also means that there are more snails reproducing when their population is threatened.

Despite their seeming unimportance, our Creator has designed snails to play an important role in maintaining the ecology. For this reason, and because He made them, He has provided snails with the ability to prosper under stress. They do not succeed by their own effort; snails do nothing more than what the Creator made them for.

The lessons we can learn from these snails are easy to speak but sometimes hard to live. First of all, if the Creator cares this much about snails, then not one of us is so unimportant that our Creator ignores us. And when life is difficult for us, our help does not lie in our own meager powers but in the power of the One who made us and sent His Son for our salvation. In commending all to Him and doing His will, we will enable Him to give us a more complete life.

Prayer: Dear Lord, since You have walked this Earth and tasted the dust of life with Your own lips, You know the problems which come with life. Teach me to depend more and more on Your sufficient power and less and less on my own powers. Amen.

Ref: "Snail's pace picks up if hunted; it gets bigger and lives longer." *Minneapolis Star Tribune*, Feb. 24, 1990.

Growing New Brain Cells

Proverbs 22:17-19
"Bow down thine ear, and hear the words of the wise, and apply thine heart unto my knowledge. For it is a pleasant thing if thou keep them within thee; they shall withal be fitted in thy lips. That thy trust may be in the LORD, I have made known to thee this day, even to thee."

Scientists who study the brain have always believed that adult creatures have all the brain cells they're ever going to have. If this is the case for humans, with the most powerful and highly evolved brains, then of course it must be true for the "lower" animals as well.

But, the actual brain structure and demonstrated intelligence of various creatures is often just the opposite of what those who believe in evolution expect.

The working cell of the brain is called the neuron. We have trillions of them in our brains, and each wire-like neuron has connections to millions of other neurons. So growing a new, working neuron and getting it properly wired into millions of other neurons is no simple business. That's why scientists have always considered it impossible for new brain cells to develop in adults.

But now researchers are investigating convincing evidence that adult songbirds not only grow new brain cells that properly wire themselves into the brain, but that these new cells enable new learning to take place. Scientists hope to discover how this takes place in the hope that the same growth of new brain cells could be brought about in humans who are suffering from brain damage. Here is yet another example of how, by their actions, intelligent scientists are denying evolution by seeking to learn how the Creator has designed His creation.

Prayer: Dear Father, I thank You that Your handiwork and power are so clearly evident in the creation. I ask that You would use this evidence to turn the hearts of those who would deny You so that they might learn of Your love to them in Christ and become witnesses for You. In Jesus' Name. Amen

Ref: Kolata, Gina. 1985. "Birds, brains, and the biology of song." *Science 85*, Dec. p. 58.

The Miracle of Photosynthesis

Genesis 1:30
"And to every beast of the earth, and to every fowl of the air, and to every thing that creepeth upon the earth, wherein there is life, I have given every green herb for meat: and it was so."

All green plants, some algae, and even some bacteria are able to make food out of nothing more than air, water, light and a few minerals. The process is called photosynthesis, and without it we would run out of food to eat as well as oxygen to breathe.

The green plant takes in the carbon dioxide that we and the animals exhale as waste and some water and, through photosynthesis, produces oxygen and a carbohydrate, and returns three-fourths of the water it originally took in for future use. Chemically what happens is that the carbon atom in the carbon dioxide is removed from the oxygen and added to one water molecule, creating a carbohydrate, which is useful to us as food. While this all sounds simple, a more detailed summary of what happens each step of the way would fill a whole page with fine print.

Evolutionists marvel at the great good luck involved in the fact that plants make useful things for us out of our waste products. We have grown in our appreciation of photosynthesis since scientists have tried to mimic the process in order to build a new kind of solar cell for use in space. While the plant converts nearly 100% of the light it receives into energy, our best human efforts have reached only 8% efficiency.

So creationists ask, if the best human minds have produced only an 8% efficiency after years of work, how could no mind at all come up with nearly 100% efficiency – no matter how much time was involved!

Prayer: Lord, if it were up to us to design our own world, the result would have been a hopeless mess that would have destroyed itself. Help people see the foolishness of thinking that no one at all made the world and the excellence of Your saving love to us. Amen.

Less Bang for the Red Shift

Genesis 1:16
"And God made two great lights; the greater light to rule the day, and the lesser light to rule the night: he made the stars also."

Scientists who reject the claim that God created the heavens and the Earth usually say that the universe is the result of some huge explosion in space billions of years ago.

If you spatter paint spots onto the surface of a deflated balloon and then begin inflating the balloon, you will have a rough idea of how astronomers picture the universe. As the balloon inflates, each paint spot moves farther away from every other paint spot. Likewise, evolutionary scientists believe that every other star is moving away from us. And like the spots on the balloon, the most distant stars are moving away at higher speeds than the closer stars.

One evidence of this, say scientists, is that light waves from distant stars are made a little longer – that is, red-shifted – by this high-speed movement. This kind of shifting is called *Doppler shifting*. Creationists have pointed out that other things could be causing this red shifting, and that it may have nothing to do with the Big Bang. While their claims have been ignored, the scientific literature now describes a new explanation for this red shifting of light – called wolf shifting – which can scientifically account for the observed red shift of light without the need for a Big Bang theory.

The millions and billions of light years that astronomers talk about and the billions of years worth of age that they assign to the universe are all based on the Big Bang. But science is now offering other explanations that may burst the Big Bang balloon!

> ***Prayer: Heavenly Father, the stars rest in Your Hands, and it is Your Word which charges them with immeasurable power. Change me and help me to realize that Your greatest show of power is not in the stars but in the power of Your Word to create saving faith in Your Son, Jesus Christ. In His Name. Amen.***

Ref: Amato, I. "Expanding a theory for shifting starlight." *Science News*, v. 136. p. 326.

Fishy Physicists

Psalm 145:15
"The eyes of all wait upon thee; and thou givest them their meat in due season."

Did you know that tadpoles, basking sharks and many types of whales all feed in the same way? A wide range of creatures feed on tiny creatures suspended in water.

Most people assume that suspension feeders have sieves in their mouths that allow the water to pass through, trapping the tiny creatures that are to be lunch. But if you have much experience with sieves, you know that they quickly become clogged, especially when trying to filter out things as tiny as the creatures suspension feeders eat. For this reason, many suspension feeders don't rely upon sieving as much as they reply on more ingenious methods of separating lunch from water.

One popular method among suspension feeders is to intercept the tiny creatures with a sticky surface. Another method sets up water currents as water passes through the gills or mouth, which causes food particles to settle out of the moving water so they can be swallowed. But perhaps herring and sardines have the most interesting system. They have special pockets just above their esophagus that are so designed that as they pass water through the gills, the flow of water causes food particles to compact in the pockets. By the simple act of "breathing," these creatures also obtain lunch.

The engineering precision and genius in these feeding systems clearly testifies not only that there is indeed a Creator, but that He provides for the needs of His creatures.

> **Prayer: Dear Heavenly Father, I cannot possibly know all of my needs, no less supply them by myself. For Jesus' sake, forgive my independence and pride and help me learn both my need and its supply from You. In His Name. Amen.**

Ref: *Scientific American*, Mar. 1990. p. 98.

The Surprising Clown Fish

Hebrews 4:16
"Let us therefore come boldly unto the throne of grace, that we may obtain mercy, and find grace to help in time of need."

The often beautiful, but always deadly sea anemone is basically a mouth and stomach, surrounded by grasping, stinging arms attached to the sea floor. When a passing fish brushes up against the arms, thousands of microscopic stinging cells in each arm fire, stunning him. Then the arms slowly start working the fish toward and into the anemone's mouth in the center of the arms.

It is this deadly environment that the clown fish calls home. The clown fish is so named for its bright colors, but the name could as well apply to many aspects of its lifestyle. The clown fish lives among the stinging arms of the anemone without harm because it coats itself with the same mucus that prevents the arms from stinging themselves. And of course, the arms provide the clown fish with a safe refuge from enemies.

Clown fish mate for life, each pair staking out its own anemone. Their offspring gradually begin to populate surrounding anemones. But if the female dies, the male will change into a female and seek a mate from among the oldest of its unmated male offspring.

Clearly the arrangement between the anemone and the clown fish was designed to be that way from the beginning. Those who think this arrangement evolved must invent an imaginative explanation of how the clown fish learned to protect itself from the stinging arms of the anemone. No, the wisdom and beauty of this relationship is characteristic of the Creator.

Prayer: Dear Lord, I thank You for the great beauty which surrounds us in the creation every day. Let Your forgiving grace be my protection in my relationships with others. Amen.

Ref: Rotman, Jeffrey L. 1984. "Living world." *Science 84.* p. 28.

Pet Your Houseplants

Romans 5:8
"But God commendeth his love toward us, in that, while we were yet sinners, Christ died for us."

Scientists have known since the 1970s that plants have feelings. Further research is showing not only more ways in which plants react to their environments, but also how plants respond to changes.

Plants react to wind, rain, and even human touch. But since they cannot change location in response to the environment, like animals and human beings, they react in other ways. These reactions seem to be controlled by a gene within the plant that is activated by wind, rain or the human hand.

What are trees like on a wind-beaten, storm-lashed coast? Do they stand straight and tall? No, they are often short and sturdy, perhaps even bent to stand against the storms. In a protected area, that same tree might have grown straight and tall. But the constant stress of wind and rain cause the tree to respond in this way. Scientists have found that the human touch on plant leaves can have the same effect as wind and rain. Over a period of three weeks, plants that were touched twice a day were shorter and sturdier and up to half as tall as plants that weren't touched. One lesson is that if you have plants that are getting tall and leggy, you should take time each day to stroke the leaves.

We should not be surprised to learn that plants respond to their surroundings – and even to loving strokes. After all, they are the creation of God who created human beings because He wanted someone to love.

> ***Prayer: Dear Father in heaven, the message throughout Your Word is that You want to love us, even though we sinned and ruined Your creation. I thank You that You sent Your Son Jesus Christ Who purchased and won me from sin, death and the devil. In His Name. Amen.***

Ref: Decker, C. 1990. "Plants under pressure: the touch that stunts." *Science News*, Feb. 24. p. 117.

Was Behemoth a Dinosaur?

Job 40:15-17
"Behold now behemoth, which I made with thee; he eateth grass as an ox. Lo now, his strength is in his loins, and his force is in the navel of his belly. He moveth his tail like a cedar: the sinews of his stones are wrapped together."

In Job 40:15-24 the Lord points Job to an animal called behemoth as an example of His great work of creation. Could this be a scriptural mention of a dinosaur?

In the Hebrew language the word translated "behemoth" basically means "beast." "Behemoth" is mentioned in Genesis 1:24 as well. But in Job we have a physical description of "behemoth." If you look up Job 40:15-24, you may find a note in your Bible suggesting that "behemoth" was a hippopotamus. But as you read this section, the description doesn't seem to be that of a hippo. In verse 17 we read, "He moves his tail like a cedar. If you've ever seen the little rope-tail of a hippopotamus, you would find it hard to compare that tail to a mighty cedar. Nor does this description fit a rhinoceros. While it is large and heavy, the rhinoceros is no match for a raging river, as verse 23 says "behemoth" is.

This is obviously a description of a large and powerful animal – an animal that seems to be larger and heavier than a hippopotamus. Many creationists have suggested that this huge creature with a tail like a giant cedar might be a giant plant-eating dinosaur like the brachiosaurus.

We do know that humans and dinosaurs existed together. We can't be sure, of course, but perhaps Job was completely familiar with at least one kind of dinosaur!

Prayer: Dear Father, I stand in awe at the power and size of some of the creatures You have made. Yet it was for me, a relatively small and powerful creature, that You sent Your Son. Help me not to look to my own strength or cleverness but always to You and Your saving grace to me through Christ Jesus. Amen.

Ref: Fish, Robin D. 1986. "Dinosaurs in the Bible?" *Bible-Science Newsletter*, Mar. p. 9.

A Mother's Love

I John 4:8
"He that loveth not knoweth not God; for God is love."

We have all seen it many times. A mother walking in a crowd with a small child loses track of her child and begins calling and searching for the little one. But can you imagine a crowd of 70 million mothers and children all looking for each other?

That's exactly what the Mexican free-tailed bat must cope with. Once its pups are born, the bat colony may have as many as 70 million individuals, with up to 40 tiny babies for every 16 square inches. Being mammals, bats nurse their young and the young bats are very aggressive about getting something to eat. So scientists simply assumed that when the mother bat returned from feeding, she would be facing millions of hungry babies. Most likely, she would simply nurse the closest and most insistent young – unable to find her own.

But studies show that when the mother bat returns from feeding, she lands near where she left her baby and begins calling to it, listening for an answer. Her youngster will call back to her, and when she thinks she has found her offspring, she sniffs it to be sure before nursing. Researchers found that in the vast majority of cases, mothers found their own young.

Scientists were very surprised to discover how powerful the mother bat's love is, as well as how intelligent bats really are. Love is but one part of life that evolution cannot explain. But we who believe that we and all things were made by a Creator who is love have no trouble explaining where love comes from.

> **Prayer: Dear Father in heaven, You meant love and not "tooth and claw" to be the rule in creation. Teach me more of Your love so that I may communicate it to others in this world where many believe that "tooth and claw" should rule. In Jesus' Name. Amen.**

Ref: "Some mothers don't forget their children." *Science 84*, June. p. 8.

A Warm-Blooded Turtle

Psalm 150:2
"Praise him for his mighty acts: praise him according to his excellent greatness."

For generations it was a rule that reptiles were cold-blooded. And turtles, being reptiles, were therefore called cold-blooded as well. Being cold-blooded, it was expected that the body temperature of turtles would be close to that of the surrounding water.

It was creation-scientist Dr. Wayne Frair who first discovered a "warm-blooded" turtle in the early 1970s. He and two other scientists published their historic research in the journal *Science* in 1972. They reported the discovery of a giant leatherback turtle weighing 920 pounds that was able to maintain a body temperature of 72° F, even when placed in a tank of 45° water. Leatherback turtles range from warm tropical waters to cold northern waters, often migrating between them.

This discovery actually shatters two false claims made by evolutionary science. First of all, it proves that reptiles can be warm-blooded and that the leatherback turtle is warm-blooded. Secondly, this is one of many examples answering the claim that scientists who believe in creation aren't real scientists. Over the years many scientists who believe in creation have made important discoveries in science.

This shouldn't surprise us, since science studies the material world that Creator made. Scientists who believe in creation are actually in a better position to learn what He has done than those who deny Him.

Prayer: Father in heaven, let my faith in You and Your work of salvation for me help my work show forth Your excellence as a witness before all. In Jesus' Name. Amen.

Ref: Discovery. *Bible-Science Newsletter*, Apr. 1973, p. 2.

The Great Wall in Space

Isaiah 34:4
"And all the host of heaven shall be dissolved, and the heavens shall be rolled together as a scroll: and all their host shall fall down, as the leaf falleth off from the vine, and as a falling fig from the fig tree."

Some Christians believe that God's words in Genesis 1:3, "Let there be light," are a biblical description of the Big Bang that some scientists say created the universe. But perhaps we Christians should be a little more careful about assuming that modern science knows very much about the origin of the universe.

Astronomers recently announced that they had discovered the largest structure yet to be found in the universe. They described the structure as a great wall made up of high concentrations of galaxies. Just to get our perspective, the average galaxy contains over 1 billion stars. The great wall contains concentrated *"clumps" of galaxies*!

This discovery delivers two apparently fatal blows to the Big Bang theory. If the universe was the result of the Big Bang, scientists would expect to find stars evenly distributed in space, not "clumped" together and certainly not built into giant structures. Second, the clumps of galaxies they found are very precisely and evenly spaced – not the kind of order that results from an explosion. One of the researchers said, "It is safe to say that we understand less than zero about the early universe."

There is another good reason for Christians not to try to find the Big Bang in Genesis. According to the Bible, it is the end of the world and the universe, not its beginning, that could more accurately be described as a "big bang." Christ Himself has completed your preparations for that day. Are you ready?

Prayer: Dear Lord, I cling to Your saving work for my preparation for the end of the world. Be with me now and prepare me to spend eternity with You. Amen.

Ref: "Galaxy clumps' may shed light on cosmic creation." *Minneapolis Star Tribune*, Feb. 23, 1990.

Could Leviathan Be a Dinosaur?

Job 41:1, 6, 30

"Canst thou draw out leviathan with an hook? or his tongue with a cord which thou lettest down? Shall the companions make a banquet of him? shall they part him among the merchants? Sharp stones are under him: he spreadeth sharp pointed things upon the mire."

The entire 41st chapter of the Book of Job describes a creature called "leviathan." As in his earlier description of "behemoth," the Lord is impressing Job with the wisdom and might of His creating power.

And "leviathan" is an impressive creature indeed! We are told that he cannot be made to serve humans. Nor would any person make the mistake of fighting "leviathan" a second time. Leviathan is mighty and graceful, yet he regards iron as straw, for iron cannot pierce his armored scales. Even his undersides have sharp, armored scales. What could this fearsome creature be?

Some translations suggest that "leviathan" is a river crocodile. But it has been pointed out that this description doesn't even come close to fitting the most fearsome crocodile. For one thing, crocodiles don't have sharp scales on their undersides. Rather, their undersides are soft and easily pierced, even with a knife. And even in Old Testament times some people made their living killing river crocodiles and making leather of their hides. In many places river crocodiles nearly became extinct because of hunting.

No, say many creationists, this is not a river crocodile, nor does it seem to be any other familiar creature. But it is possible that it could be a water-dwelling dinosaur, perhaps like a tylosaurus. We can't be sure, of course. But the creature described in Job 41 does fit what we know about these giant reptiles.

Prayer: Lord, there are many surprises for us as we discover new examples of Your power and creativity. Keep my mind open to learning these new things so that I may have yet more reason to praise and glorify You. Amen.

Science Looks at Astrology

Isaiah 47:13
"Thou art wearied in the multitude of thy counsels. Let now the astrologers, the stargazers, and the monthly prognosticators, stand up, and save thee from these things that shall come upon thee."

Judging by the number of newspapers that carry daily horoscope columns, astrology continues to be very popular. Unfortunately, even many Christians have more than a passing interest in astrology.

Interest in astrology goes back thousands of years. And so do God's warnings for His people to stay away from astrology. However, it is only within the last 30 years that science has subjected astrology's claims to careful testing. Recent studies have turned up all sorts of problems with astrology. For example, as the Earth rotates, it also wobbles. As a result of that wobble, the zodiac upon which astrology is based has changed by nearly one entire constellation in only the last 2,000 years. Astrologers have never accounted for that change.

Likewise, studies have shown that horoscopes themselves are inaccurate. A study of men reenlisting in the Marine Corps from 1962 through 1970 showed that their astrological signs were just as likely to be ruled by Venus, the planet of love, as they were by Mars, the god of war. In a study of nearly 3,000 couples, researchers found that astrologically incompatible signs were no more likely to be divorced than astrologically compatible signs.

The Bible opposes astrology for a more important reason than that it is false science. God warns His people away from astrology because it encourages us to trust in created things, like the stars, instead of the true God who made us and loves us.

Prayer: Dear Father in heaven, so many created things ask for my trust, a trust that belongs only in You. When I am tempted to place my trust in something created, let me see what I am doing and cling once again only to my Lord and Savior Jesus Christ for forgiveness. In His Name. Amen.

Ref: Waldrop, M. Mitchell. 1984. "Astrology's off target." *Science 84*, June. p. 80.

Even Bacteria Get Sick

Romans 8:22
"For we know that the whole creation groaneth and travaileth in pain together until now."

It may come as good news to learn that the bacteria that make us sick can get sick too. Somehow it seems only right that they get a little of their own medicine, so to speak. And it may well be that making disease-causing bacteria sick is just the medicine we need.

The fact that bacteria have their own microscopic enemies was first discovered in 1925. Antibiotics were far in the future, so scientists began to learn about the bacteria's enemies in order to start some real germ warfare against other germs. They hoped their efforts would lead to cures for pneumonia, tuberculosis, cholera and diphtheria. But no one was able to perfect a treatment that worked.

Today our much more advanced medical research has again become interested in this subject, since more bacteria seem to resist our best antibiotics. Research is centering on viruses called bacteriophages, which literally means "bacteria eaters." An hour after a bacteriophage has infected a bacterium and reproduced within it, it kills that bacteria and moves on to others. In recent tests, bacteriophages have proven more effective than antibiotics in curing some livestock infections. And since they are alive, bacteriophages are passed from one animal to another, sharing the disease resistance.

Once again we are learning from our Creator how to ease human suffering.

> ***Prayer: I thank You, Lord, that You enable mankind to learn how You have made things so that we have less suffering in this life. Fill people with an increased desire to know how our deeper suffering, because of sin, is cured through the saving work of Christ. Amen.***

Ref: Dixon, Bernard. 1984. "Attack of the Phages." *Science 84*, June. p. 66.

Tent-Building Bats

Psalm 14:1
"The fool hath said in his heart, There is no God. They are corrupt, they have done abominable works, there is none that doeth good."

Have you ever noticed that many who believe in evolution depict early humans as so primitive that they had to live in caves while many kinds of animals build much more sophisticated shelters? This inconsistent thinking is built on the evolutionary assumption that intelligence, like everything else, has evolved. Creatures that evolutionists think evolved much earlier must be less intelligent.

It was because of this unscientific evolutionary view that the scientist who first discovered the South American bats who build and live in tents refused to believe the bats were normal. The "tent-making" bat, as it is now called, carefully cuts the veins in palm leaves and folds the leaves over to create a protective tent. Each cut is made as though the bat is following a blueprint, and the resulting tent completely hides the bat within what appears to be nothing more than a folded over leaf.

We now know that fourteen New World bat species, as well as two Old World species, build tents. Scientists have also learned that the various species of tent-building bats have complex social customs that determine the design of the tent and how many individuals are allowed to live in one tent.

Tent-making bats offer yet one more example of how the living world rises far above the expectations of those who believe in evolution. The limitations of human explanations should remind us that our focus should be on the things of God if we truly want to learn what we need to know in the world.

Prayer: Dear Lord, You have allowed man to make many accomplishments. But do not let Your people be fooled into thinking that man is able to discover meaningful answers about his existence on his own. Increase the voices You have sent to remind us to look to and study Your Word. Amen.

Ref: Timm, Robert M., and Barbara L. Clauson. 1990. "A roof over their feet." *Natural History*, Mar. p. 55.

Breaking Dollo's Law

Proverbs 28:1
"The wicked flee when no man pursueth: but the righteous are bold as a lion."

Do evolutionists believe that life always becomes more complex and sophisticated? Does evolution ever go backward?

In 1893, a man named Dollo proposed a law that has become a cornerstone of evolutionary belief. Dollo's law says that evolution always goes uphill toward more specialization. It never goes backward. This means that the higher we go up the evolutionary ladder of development, the more development we should see in at least one, if not more, parts of a creature. On the other hand, creationism says that the Creator made each kind of animal distinct for different reasons. Because of this, so-called lower animals might have more highly developed cells or organs than so-called higher animals.

Now let's test these two conflicting claims and see which one fits what we find in the real world. Which view is supported in studies comparing the muscle tissue of the horseshoe crab, a very early creature according to evolution, with the rabbit, a comparatively late creature? Studies show that the protein structure of horseshoe crab muscles is more complex than in rabbits, or in many instances, even humans.

These facts should help Christians feel a little less intimidated by evolutionists who claim that evolution is a fact of science. One of the most basic laws of evolution fails to stand up to testing.

Prayer: Dear Father, I confess that I am too easily intimidated by humans values like educational degrees. Forgive me for the sake of Jesus Christ and what He has done for me, and fortify me with the boldness of faith to speak a clear witness to Your truth. In Jesus' Name. Amen.

Ant Antics!

Proverbs 30:25
"The ants are a people not strong, yet they prepare their meat in the summer;"

Proverbs 30:25 uses the word "people" in referring to ants. In the Hebrew, this word is the very same word commonly used throughout the Old Testament for people, including the people of Israel. While this might at first seem strange, it is true that in many respects ants do seem like tiny people.

Ants love to work and live in sunlight. They will spend hours clearing their little backyards of leaves and even plants and shrubs. Ants are also known to wander far from their local area. One scientist from the University of California tagged ants with colored dyes and then watched their meanderings. No matter how far they wandered, they didn't seem to get lost. Then the scientist placed tiny ant blinders on the wandering ants. He discovered that when the ants couldn't see their surroundings, they wandered aimlessly, obviously lost. This and other experiments have shown that ants have incredible memories.

Scientists have also marveled at the tricks used by ants to ruin picnics. One of the ants' most interesting strategies is to climb a tree over a picnic and find a low-hanging leaf situated over the picnic basket. The ant then proceeds to chew off the stem of the leaf and glide down into the sandwiches!

Read Proverbs 6:6 and 30:25 and see for yourself that Scripture is true, even when it talks about the world that science studies – and even about such details as ants!

Prayer: *I thank You, Lord, that Your Word is absolutely trustworthy in all that it says. Thank You for preserving the Bible for thousands of years so that I and millions of others may learn of Your plan of salvation for us. Amen.*

Engineering Joint Lubrication

Joshua 24:15
"And if it seem evil unto you to serve the LORD, choose you this day whom ye will serve . . . but as for me and my house, we will serve the LORD."

In our rapidly modernizing world, engineers are kept busy solving problems. Take for example, all of the various kinds of transportation. There are millions of problems in this area alone that keep engineers busy inventing better solutions.

Freight trains carry enormous loads in huge freight cars, each one capable of carrying the weight of the average home and everything in it, plus an automobile. Yet the axles must be able to swivel easily beneath the car as the train moves along tracks that swerve left and right. Consider the problem of setting 100 tons or more on a swivel without hindering the free movement of the swivel. Engineers at Shell Oil finally designed a disc that is placed beneath the body of the car to lubricate the axles for swiveling. Whenever the disc is squeezed by the weight of the load above or from too much friction in swiveling, lubrication automatically squirts out of the disc.

It is this same ingenious system that lubricates certain joints in your body. When additional lubrication is needed in a joint, tiny discs release lubricant into the joint. These discs are called bursae. And if you have ever had bursitis, you know what happens when the discs aren't working properly.

It's not science but faith in evolution that leads people to believe that this well-engineered system could be the result of accidental mutations.

Prayer: *I thank You, dear Lord, that You have designed my body in such a wondrous fashion. Help me to take good care of it and grant me good health and strength to be used in service to You. Amen.*

The Bare Bone Facts

Psalm 139:13-14
"For thou hast possessed my reins: thou hast covered me in my mother's womb. I will praise thee; for I am fearfully and wonderfully made: marvelous are thy works; and that my soul knoweth right well."

When an engineer builds a building, a bridge, or some other structure, he must build it so that it can withstand both stretching and compressing forces. In designing the structure to withstand both kinds of these forces, he must anticipate how much of each force the structure might face in its lifetime.

Some materials, like cast iron and concrete, are very good at withstanding compression forces but very poor in handling stretching forces. Using one of these materials to build a structure that undergoes tensile or stretching forces could lead to disaster.

Now let's apply our basic engineering information to the problem of building a frame or skeleton for a living creature. Although you may weigh only 130 pounds, your long leg bone will very likely have to be prepared to deal with more than 1,000 pounds of compression and hundreds of pounds of tension from the muscles that are anchored to it. It's a good thing, then, that normal bone is three times as strong as good solid wood and nearly as strong as iron! Tests have shown that the tensile strength of bone is 35,000 pounds per square inch, while iron is 40,000 pounds per square inch. But bone material is better than iron because it is three times lighter and much more flexible.

The structure of your bones is too carefully engineered to have been an accident. If life was the result of impersonal evolution, evolution would still be trying to engineer the best material for skeletons, and there would be an awful lot of jellyfish around!

Prayer: Dear Father in heaven, help me to remember that You also care about the spiritual stresses in my life and that You are present to help me, as Your redeemed child in Christ, if I will but take my needs to You. Thank You for being there when I need You. Help me to need You always. In Jesus' Name. Amen.

Natural Acid Rain

Isaiah 51:6
"Lift up your eyes to the heavens, and look upon the earth beneath: for the heavens shall vanish away like smoke, and the earth shall wax old like a garment, and they that dwell therein shall die in like manner: but my salvation shall be for ever, and my righteousness shall not be abolished."

One big issue these days is acid rain falling in industrialized areas of the world. Much of the acid in rain comes from nitric acid and sulfuric acid, which are released as a result of automobile exhaust and factory smokestacks. While rainwater needs to be somewhat acidic to dissolve minerals so they can be taken up by plant roots, too much acid is damaging buildings, automobiles, forests and lakes in some areas of the world.

Since acids tend to wash out of the atmosphere in rain fairly close to where they are created, scientists were amazed to find acid rain contamination in remote areas of the Amazon River basin, far from any cities. They identified the acid as formic acid, byproduct of formaldehyde, a common industrial exhaust pollutant.

After more research, scientists discovered that the Amazon acid rain problem was being caused by ants! There is a huge population of formicine ants in the Amazon basin. These ants produce formic acid, which they store in a pouch and use both to defend themselves and to communicate with each other. Each ant might carry enough formic acid to make up five percent of its weight, which doesn't seem like very much. But when you multiply that by 100 trillion ants, you have enough formic acid to create an industrial-sized pollution problem!

There are indeed problems on this earth. The Bible warns us that the earth is wearing out like a garment. Our Creator will soon come for us and then there will be a new heaven and a new earth!

Prayer: Dear Lord, I look forward to Your return to take me and all believers to Yourself forever. By Your Holy Spirit, enable me to make every day of my life a day of preparation for You. Amen.

Ref: "100 trillion ants drop acid." *Discover*, Sept. 1987. p. #8.

Ant Mathematics

Luke 14:28-30
"For which of you, intending to build a tower, sitteth not down first, and counteth the cost, whether he have sufficient to finish it? Lest haply, after he hath laid the foundation, and is not able to finish it, all that behold it begin to mock him, Saying, This man began to build, and was not able to finish."

Can ants count? It seems so! When scout ants find an item of food, they take it back to the nest. If the food item is especially good but too big to carry, the scout will return to the nest to get help. Scientists have discovered that ants apparently size up the task ahead before getting help so they can return with enough help, but not too much.

One scientist cut a dead grasshopper into three pieces. The second piece was twice the size of the first, and the third was twice the size of the second. He then left the pieces in different locations where ants were sure to find them. He watched as each piece was discovered by a scout, inspected, and each scout returned to the nest for help. When the scout returned with help, the scientist counted the number of ants working at each piece of the grasshopper.

The smallest piece had 28 ants working on it. The piece that was twice its size had 44 ants working on it. And how many ants do you think were working on the piece that was twice the size of the second piece? If you doubled that 44 to 88, you would be within one of being right – there were 89 ants working to return it to the nest!

We can't help but conclude that mathematical ability is part of the ants' amazing ability to plan and carry out a task! They were doing nothing more than what their Creator taught them as they followed the planning principle Jesus reminds us of in Luke 14:28 – count the cost of the project before you begin it!

Prayer: Dear Father, I confess that too often I have planned my life, leaving You out of my planning. But who knows more about planning than You? Grant me wisdom and guidance so that my life may be pleasing to You in every way. In Jesus' Name. Amen.

Yes, Early Humans Wrote

Exodus 17:14
"And the LORD said unto Moses, Write this for a memorial in a book, and rehearse it in the ears of Joshua: for I will utterly put out the remembrance of Amalek from under heaven."

Evolutionary scientists have been amazed by the discovery that the *very oldest* artifacts left by humans indicate written language and mathematical ability. Creationists, on the other hand, are as pleased as can be at the findings, since they predicted that it would be possible to find that humans have had language throughout their history on Earth. After all, they were created by the word of God – who later took human flesh upon Himself – and before the Fall away from God into sin, humans were able to speak directly with God.

While we do not accept the inflated evolutionary years, evolutionists now say that human written language goes back at least 30,000 years. Not long ago they said writing had not even been invented at the time of Moses!

When archaeologist Alexander Marshack determined that one ancient carving was a calendar, scientists sat up and took notice. This discovery meant not only that ancient humans had a sense of time, but that they also had a written language and the desire and ability to do math. No matter how far back one is able to find evidence of humanity, language was already established.

Humans have always been human. If our actions are less than human, it is because sin has taken over. But there is rescue from sin in Jesus Christ. If you don't know Him, or if your life seems less than human, take a look at the Bible and see what He has to offer you.

Prayer: Dear Heavenly Father, because of his sin, man seeks escape from You. He is so driven to escape that he would rather see himself as a child of apes than an a child of God. Help me to bear witness to Your love for us in Christ so that those around me desire Your love more than escape. In Jesus' Name. Amen.

Ref: "Ice age artists open doors on early man." *Science Digest*, May 1978. p. 87.

The Wonders of Everyday Materials

Isaiah 45:18
"For thus saith the LORD that created the heavens; God himself that formed the earth and made it; he hath established it, he created it not in vain, he formed it to be inhabited: I am the LORD; and there is none else."

Today lasers do many important jobs in manufacturing and medicine, in addition to serving in weapons that once existed only in the minds of science fiction writers. But if our small imaginations can think of amazing things, we shouldn't be surprised to learn that our Creator's limitless imagination can think of even more incredible things. As science studies the handiwork of the Creator, we are learning about His imagination.

For example, scientists have recently discovered how to make hydrogen – the simplest of all elements – into a solid. Hydrogen is technically a metal, even though it is a gas at most temperatures. Yet scientists have expressed amazement at the complexity of this so-called simple element. When compressed under 1.5 million atmospheres, near absolute zero, hydrogen becomes a solid metal – clear as glass!

Many people do not know that glass is really a liquid at normal temperatures. Your windows are even now flowing downward because of gravity. It's just happening so slowly that you would have to compare photographs of the same window taken a century apart to see the effect. And now scientists are working on metal films that are transparent to light. Perhaps someday we'll look out through windows of aluminum instead of glass!

Such amazing discoveries about seemingly simple, everyday materials show us that while humans may have imagination and inventiveness, our Creator has even more imagination than we could ever have!

Prayer: Dear Lord, nothing is too hard for You. The wonderful surprises we are finding in Your creation provide yet more witness to modern man that You, and not mindless chance, have designed everything in the material world. Let this witness draw more of our modern world to You and Your love for us in Christ. Amen.

Ref: "Fashioning see-through metal." *Science News*, July 8, 1989. p. 31.

Why the Boomerang Returns

Psalm 119:73
"Thy hands have made me and fashioned me: give me understanding, that I may learn thy commandments."

For many people the fun of the boomerang lies in the fact that it seems to go against common sense. You throw the boomerang away from you and it returns to you.

Boomerangs of various shapes and sizes and made of different kinds of materials are easily available in many places around the world. Boomerangs can be made into many shapes other than the classic "L" shape we usually think of. They have been made into the shape of a windmill with six blades. Boomerangs have also been made into the shapes of twenty letters of the English alphabet. Contrary to popular myth, boomerangs never were used as weapons by Australian aborigines. They used them as toys just as we do today.

What makes the boomerang return to the thrower? The arms of the boomerang are curved on top and flat on the bottom, like an airplane wing. So as the boomerang moves through the air, its wings create lift just like an airplane's wings. But as it spins, the edge of the boomerang wing that is speeding forward in spin creates more lift than the relatively slower opposite arm. This causes the boomerang to tip. As a result of a principle known as gyroscopic precession, this uneven lift causes a constant turning pressure on the boomerang, eventually returning it to the thrower.

The boomerang uses sophisticated aerodynamic and physical principles. It clearly was not invented by people who were primitive. Rather, it is a testimony to the intelligence that the Creator gave to every human being – from the first to the last.

Prayer: Dear Lord, I ask that You would help me to make good use of the intelligence You have given me so that I may serve you more effectively and more joyfully. Amen.

Ref: Robson, David. 1983. "Many happy returns." *Science 83*, Mar. p. 100.

When It's Better to Be Male

Matthew 6:30
"Wherefore, if God so clothe the grass of the field, which to day is, and to morrow is cast into the oven, shall he not much more clothe you, O ye of little faith?"

Only recently have botanists begun to appreciate the jack-in-the-pulpit. This common flowering plant, which lives from 15 to 20 years, is found throughout eastern North America.

Middle-sized jack-in-the-pulpits normally have only one leaf and are male. The male jack-in-the-pulpit's primary job is to produce pollen to fertilize female plants. The female jack-in-the-pulpit is larger and usually has two leaves besides its flower, which, when fertilized, produces seeds. Smaller plants have no flowers at all and are neuter until they grow larger.

Botanists have long known that the jack-in-the-pulpit changes sex. A male plant producing pollen this year may be a female plant producing seeds next year. Now botanists are discovering that jack-in-the-pulpits can change sex fairly often, and they are learning why. It requires much more energy for the female jack-in-the-pulpit to produce seeds than it does for the male to produce pollen, especially since jack-in-the-pulpit seeds are unusually large. If the weather is poor or a female jack-in-the-pulpit uses too much of her energy producing seeds, she simply retires from being a female and spends the next year or two living the comparatively easy life of a male.

The jack-in-the-pulpit must give headaches to evolutionists who cannot explain how male and female could evolve in the first place. At the same time, this interesting flower annually shows how the Creator cares for everything in His creation. Learn more about His very special love for you in the pages of the Bible.

Prayer: Dear Lord Jesus Christ, there is no end to God's love. I thank You for Your saving work on the cross and Your Resurrection for my salvation. Do not let fear cause me to wander from my faith in Your daily, personal love. Amen.

Ref: Batten, Mary. 1983. "Sex-in-the-pulpit." *Science 83*, March. p. 85.

The Body's Fleeting Workers

I Corinthians 12:18
"But now hath God set the members every one of them in the body, as it hath pleased him."

Inside your body there is a large and amazing family of chemical workers who, although they usually last less than a minute, make life possible. There are so many different kinds of these chemicals, called prostaglandins, that science is just beginning to learn how important they are to life.

Prostaglandins are made by just about every tissue in your body. They are made by tissue cells from stored fatty acids. When triggered, fatty acids swarm out of the cell walls and are quickly changed into the necessary prostaglandin. Prostaglandins are involved in regulating reproduction, breathing and circulation, among other things.

Prostaglandins made by the cells lining our blood vessels relax the muscles around them so that more blood can flow through them. And blood platelets also produce another prostaglandin; when triggered, it enables the blood to clot and seal wounds. In the lungs, prostaglandins regulate the openings of air passages. They help protect the inside of your stomach. It is because aspirin inhibits prostaglandin production that it helps headaches and can cause stomach problems in some people. Prostaglandins have been implicated in the swollen, painful joints caused by arthritis.

The human body is literally a symphony of thousands of carefully designed systems, each one playing in harmony with the other and all of them working together. This fact alone leaves no justification for the claim that we are designed by genetic accidents.

Prayer: *Lord, I thank You that I am fearfully and wonderfully made. I understand that it is because of sin that sometimes our systems do not work as You designed them. According to Your will, grant me healing in those cases, but let me always be aware of Your love and presence in my life. Amen.*

Ref: Shodell, Michael. 1983. "The prostaglandin connection." *Science 83*. p. 78.

Electric Avalanche

Matthew 24:27
"For as the lighting cometh out of the east, and shineth even unto the west; so shall also the coming of the Son of man be."

Scientists have found that the usual explanation for lightning is not quite true. School children are taught that as a cloud moves through the air it picks up electrical charge because of turbulence. This eventually causes lightning.

But scientists have found that thunderclouds are able to generate only one-tenth of the electrical field that is necessary to create a lightning bolt. Without something else, there would be no lightning on Earth.

That something else turns out to be cosmic rays. Think of electricity as electrons flowing like water. While the cloud doesn't generate enough electrons to cause a flow of lightning, cosmic rays penetrating the cloud detach additional electrons from the oxygen and nitrogen in the cloud. These extra electrons are then accelerated in the cloud's electric fields, producing small avalanches of electrons. These avalanches grow and combine as they flow to the bottom of the cloud, finally to erupt as a stroke of lightning!

The Bible tells us that Jesus Christ is returning to Earth to judge all mankind. When He does, says the Bible, everyone will know about it because His return will be like a giant bolt of lightning crossing the entire sky. And just as the time to make yourself safe from lightning is before it strikes, so the time to find out how to be prepared to meet Christ is before He appears. The details of how He has made these preparations for you are found in plain and simple language in Scripture – which tells of His work of salvation. Begin your preparations now!

Prayer: Lord Jesus Christ, I eagerly await Your return to Earth. Yet I know many people who are not ready for You because they do not trust in Your saving work for them. Use me to help them trust only in Your saving work for their salvation. Amen.

Ref: Lampe, David. 1978. "Lightning research strikes a windfall." *Science Digest*, Apr. p. 25

Aspirin

Psalm 25:17-18
"The troubles of my heart are enlarged: O bring thou me out of my distresses. Look upon mine affliction and my pain; and forgive all my sins."

The aspirin tablet has been hailed as the miracle drug of the 20th century. Although it took the 20th century to flavor, coat and stamp little letters on aspirin, people have been using aspirin for centuries. The most common form in which aspirin has been taken is as willow bark tea.

Aspirin has relieved billions of headaches and brought nearly as many fevers under control over the years. In recent years aspirin has become even more important since studies have shown that as little as half an aspirin per day can help prevent heart attacks.

Despite its long-standing use in medicine, physicians had no idea how aspirin worked until a few years ago. Researchers now know that aspirin does not simply mask the pain of a headache, but it chemically turns off the cause of the average headache. It does this by deactivating the chemical made by the blood vessels that causes them to constrict. Another member of that same family of chemicals, called prostaglandins, prevents blood platelets from sticking together, helping blood to flow more freely and preventing heart attacks. This is also why aspirin can cause stomach bleeding.

In His foreknowledge our Creator knew, regretfully, that the human race would fall into sin. He knew that among the effects of a world changed by sin would be pain and sickness. So in His mercy He provided plants that would make the active ingredient in aspirin. But the greatest expression of His love for us was in providing us with a cure for sin itself in His Son, Jesus Christ.

Prayer: Dear Heavenly Father, I give You praise and thanksgiving for Your generous and undeserved mercy which provides for both spiritual need of peace with You as well as our material needs here in this world. In Jesus' Name. Amen.

Ref: Shodell, Michael. 1983. "The prostaglandin connection." *Science 83*. p. 78.

Mindless Logic?

Romans 1:20-22
"For the invisible things of him from the creation of the world are clearly seen, being understood by the things that are made, even his eternal power and Godhead; so that they are without excuse: Because that, when they knew God, they glorified him not as God, neither were thankful; but became vain in their imaginations, and their foolish heart was darkened. Professing themselves to be wise, they became fools,"

Do you and I think? Or do we just think that we think? If you believe that we were created by God, your answer to that question will be that we actually do think. But if you believe that life is just a chemical accident, you might agree with those evolutionists who actually believe that we do not think. We only think we think.

Do you have a mind? What is that part inside your body that is really you? For thousands of years we referred to this part of us as the soul. But today, with the strong humanistic desire to remove all spiritual references from language, many people simply call it the mind. The problem (for those who believe in evolution) is that saying that we have a mind still sounds a lot like saying that we have a non-material part to our being. Besides, evolutionists generally admit that even evolution could not accidentally create something as powerfully intelligent as the mind. As one evolutionist put it, there is no central intelligence within us that is in charge, because nothing could be that smart.

It seems to be a contradiction, then, that some evolutionists have written entire books explaining that none of us really has a conscious mind. One wonders why people who supposedly have no conscious minds would want to read a book written by someone else who also had no conscious mind.

Romans 1 tells us that when people deny their Creator, He eventually lets them sink into complete foolishness. As the scriptures declare: "Claiming to be wise they became fools."

> **Prayer: *Dear Father, none of us is perfect and so some foolishness even creeps into my life. I ask that You would make me eager to correct that foolishness with Your good and wise instruction in Scripture. In Jesus' Name. Amen.***

Ref: Hoffman, Paul. 1987. "Your mindless brain." *Discover*, Sept. p. 84.

When Facts Aren't Facts

II Timothy 3:14-15
"But continue thou in the things which thou hast learned and hast been assured of, knowing of whom thou hast learned them; And that from a child thou hast known the holy scriptures, which are able to make thee wise unto salvation through faith which is in Christ Jesus."

Canadian and U.S. papers have been filled with the results of recent polls in both countries that tested members of the public by asking some scientific questions. The problem is, both polls were heavily stacked with questions that resulted in so-called "wrong" answers if the respondent didn't believe in evolution. Since more than 10% of the questions dealt with evolution, we can assume that the real purpose of the "poll" was to make belief in evolution look the same as "scientific literacy."

Consider the questions asked. Of the 14 questions, one read; "Human beings as we know them today developed from earlier groups of animals." Obviously the so-called "correct" answer is "true." Another question asks: "The earliest humans lived at the same time as dinosaurs." And the "correct" answer to this is "false."

The poll does show widespread ignorance of science in real areas of science. A scientist who didn't know that light travels faster than sound would be considered unfit to do science. Measurements of the speed of light and the speed of sound are easily done and offer consistent results.

But no one has ever measured – no less reproduced – the results that the poll claims are "correct" about evolution! Many scientists, including some who are very famous, would also have insisted on the so-called "wrong" answers about evolution! The lesson for us is that, as Christians, we need to read and listen very carefully to what is called "news," because some of it really isn't based upon fact at all.

Prayer: Dear Lord, I thank You that unlike man's writings Your Word is absolutely perfect and accurate in everything it tells me. Help me to remember that the purpose of Your Word is to make me wise unto salvation, which is in Jesus Christ. Amen.

Ref: "Read it and weep." *The Globe and Mail* (Toronto), Mar. 1, 1990. p. A7.

Birds Who Build Pyramids

Job 12:7
"But ask now the beasts, and they shall teach thee; and the fowls of the air, and they shall tell thee."

Bee eaters are birds whose way of life and behavior are both intelligent and unusual. There are 24 species of bee eaters.

Bee eaters make their living catching and eating bees and wasps with stingers. The poison in many of these stinging insects is powerful enough to kill bee eaters, But the birds are not only skilled at avoiding stings; they know how to remove the poison from the bee when they eat it. Having captured a bee or wasp, a bee eater will take it to a branch where he will pound its head and rub its stinging end until all of the poison has been removed from the insect's venom sac. Once the poison is removed, the bee eater enjoys lunch.

Bee eaters are described as lively and sociable. You seldom see one roosting all by itself. And when the weather is cool, bee eaters huddle together to keep each other warm. There are even reports that bee eaters will roost on each other's backs, forming a feathered pyramid made out of birds.

Now it's possible that bee eaters figured out that they were warmer when huddled together, although even that much intelligence had to come from their Creator. But how could bee eaters simply "discover" how to detoxify bees? If this ability evolved by trial and error, there would probably be no descendants of the first bee eaters around today. Obviously this dangerous behavior would not favor survival. This makes the bee eater one of God's own arguments against evolution!

Prayer: Dear Heavenly Father, not only does Your wisdom surround us, but You have so generously given intelligence and wisdom to so many of Your creatures. I thank You for the wonder Your handiwork inspires. In Jesus' Name. Amen.

Ref: Clanbake. *Natural History*, Mar. 1990. p. 94.

Love with Eight Arms

Matthew 23:37
"O Jerusalem, Jerusalem, thou that killest the prophets and stonest them which are sent unto thee, how often would I have gathered thy children together, even as a hen gathereth her chickens under her wings, and ye would not!"

While many people think that the octopus is an ugly creature, the octopus is intelligent. The mother octopus puts a lot of love and effort into the care of her young.

Fertilized octopus eggs are about the size of a grain of rice. They are connected in clusters on slender stalks. Once her eggs are ejected and fertilized, the mother octopus hangs the threads containing her eggs from the roof of her home. It could be a rock cave or even a dark, enclosed man-made structure. By the time she is done hanging her eggs, she will have almost 20,000 ivory capsules strung around her.

But now her real work begins. Until the eggs hatch, mother octopus will give up food, concentrating on nothing but cleaning her eggs of any material that might result in a fungus or parasite infection. She cradles the eggs gently in her arms, brushing each of the 20,000 eggs, and sometimes washing them with jets of water. There is nothing else she can do for her young, but this care receives her single-minded attention.

It is rightfully said that one cannot create something he doesn't have. If one does not have love, he cannot create love in others. Therefore the Creator of the octopus must know a great deal about giving individual love and care to a multitude of individuals. The Bible is our Creator's own personal description of His love for you and me. Whether you know the personal care and love of your Creator or you don't, you can learn the depth of His care for you through Jesus Christ in the pages of the Bible.

Prayer: Dear Father, not only is Your skill and wisdom evident in creation, but we can also see hints of Your love as well. Help me to realize more completely what Your personal and intimate love for me really means in my day-to-day life. In Jesus' Name. Amen.

Ref: Ruggieri, George D., with Norman David Rosenberg. 1978. "The healing sea." *Science Digest*, Aug. p. 18.

The Aardwolf

Job 15:8
"Hast thou heard the secret of God? and dost thou restrain wisdom to thyself?"

It looks something like a small hyena, has habits similar to both wolves and the domestic cat, and eats almost nothing but ants. This strange creature is called the aardwolf.

The aardwolf lives on the African continent. It eats a particular type of termite that is about one-fourth of an inch long. Although there are other ant-eating creatures in the same habitat, like the aardvark, they don't prefer to eat those termites upon which the aardwolf depends. These termites build dome-shaped mounds guarded by soldier termites that squirt long threads of nasty smelling and bad tasting chemicals at intruders. The aardwolf seems to tolerate their attacks better than any other ant-eating creature, although, as one naturalist put it, the aardwolf clearly does not like the taste of the soldier termites. Both the ants and the aardwolf are active at night. In one night the aardwolf may eat up to 300,000 termites.

The aardwolf is strongly territorial, tolerating no competitors in its territory. As solitary creatures, the mates will share the same territory yet completely ignore each other. But when it is time for mating and raising the young, they work and live together perfectly, the male helping the female in every aspect of raising the young. And, like the domestic cat, the aardwolf buries its waste.

Evolutionists admit puzzlement about how the aardwolf is related to other creatures. This makes the aardwolf yet another of those creatures that witness to God's inventive creativity.

Prayer: Dear Lord, as I look in wonder at all that You have made, I thank You that there is no limit to Your creativity. Help me to remember this wonder when I find myself in a situation where I fail to see how You are working for my good. In Jesus' Name. Amen.

Ref: Richardson, Philip R.K. 1990. "The lick of the aardwolf." *Natural History*, Apr. p. 78.

The Deep Diving Leatherback

II Corinthians 4:6
"For God, who commanded the light to shine out of darkness, hath shined in our hearts, to give the light of the knowledge of the glory of God in the face of Jesus Christ."

The fact that they must breathe air would seem to limit sea creatures like whales, dolphins and sea turtles. But these creatures are so well designed that their abilities amaze even the most informed scientists.

How deep do you think a seagoing air breather can dive while holding its breath? Keep in mind that as the creature descends through the depths, the air supply in its lungs becomes compressed and will not last as long as breath held near the surface. Until recently, whales have held the record. Sperm whales have been documented as deep as 3,500 feet beneath the waves.

Recently a tagged leatherback turtle went down to at least 3,900 feet, nearly three-fourths of a mile! According to the scientists involved in the study, the turtle might have gone deeper, but at 3,900 feet their measuring device went off the scale. In other words, they never expected the leatherback to dive so deep!

While all of us are in awe over the ability of the leatherback turtle, those who believe in evolution have a special reason to be surprised at what they are learning. According to evolution, whales and sea turtles evolved from creatures who live on the land. Evolutionists have a difficult time accounting for the extensive specialized changes required to turn a land turtle into a sea turtle capable of operating beneath three-fourths of a mile of water. Those of us who know that all things were created by a wise Creator understand that there is nothing He cannot do.

Prayer: Lord, I thank You that the whole creation shows forth Your glory and handiwork. I pray that more people, especially scientists, would see and understand this witness and be drawn to Your message of forgiveness in the Bible. In Jesus' Name. Amen.

Ref: Yertle:1, Orca: 0. *Discover*, Sept. 1987. p. 14.

Are There Black Holes?

Acts 1:11
"Which also said, Ye men of Galilee, why stand ye gazing up into heaven? this same Jesus, which is taken up from you into heaven, shall so come in like manner as ye have seen him go into heaven."

A black hole is described as an object in space so massive that its own gravity has caused it to collapse in on itself until it is immeasurably small. This increases its gravity to the point where the gravity even pulls light into its center. At this point the object has effectively disappeared from the known universe and even the normal laws of space and time don't operate. However, it's still there, its incredible gravity sucking in everything in space that gets too close, including light.

But the question is, is there any such thing as a black hole? The idea of black holes grew out of Einstein's theory of relativity. Obviously no one has ever seen a black hole. But scientists who are looking for them believe they would cause some effects in space around them that could be seen. Scientists have been searching for evidence of a huge star gravitationally collapsing in on itself. They know that gravitational collapse does happen. But they don't know if a super-giant star could collapse in this way, or if it would result in a black hole.

So far, black holes are simply theory. Not all scientists are convinced that every prediction made by Einstein's theory of relativity is accurate. But whether or not there are black holes, we do know that one day the entire universe as we know it shall end when our Lord returns to begin the new heavens and new earth.

__Prayer: Dear Lord Jesus Christ, prepare me now through Your Word so that I will be ready for You when You return for me. Amen.__

The Bats Who Feed Trees

Psalm 104:27-28
"These wait all upon thee; that thou mayest give them their meat in due season. That thou give them they gather: thou openest thine hand, they are filled with good."

The surface of the Earth 50 to 150 feet below the great living canopy of the rain forest is a dark, humid, still world dominated by great columns of tree trunks. Within those trunks and some of the giant hollow branches extending from them lies the secret of the life of the canopy itself.

A rain forest typically receives 12 or more feet of rain per year. That much rain washes out of the soil most of the nitrates needed by the trees for growth. What the trees need is a rich source of nitrates that is constantly being replaced. And that's where the bats come in.

Typically these huge trees are hollow inside. Many different kinds of creatures, including fruit bats, enter the hollow trees through various openings. Fruit bats find ideal daytime sleeping rooms inside the great hollow branches that extend like caverns from the hollow tree trunk. The accumulating layer of bat guano inside the tree itself is one of the richest sources of nitrates known. So the bats provide the tree with the nitrates it needs in exchange for a protected home during the day!

In a very real sense the fruit bats are the collection and transport system for the raw materials that make possible the tropical rain forest canopy with its millions of residents. The Creator has devised an ingenious way to provide for the needs of many creatures. Truly the Lord does provide all living things with their food in just the way they need it!

Prayer: Dear Lord, I pray that You would fill me with as much love for the creatures of Your wonderful creation as You have. And just as the entire rain forest must look to the fruit of the bat for life, help me always to look to my Savior, Jesus Christ, for all life. Amen.

Ref: Perry, Donald. "Life in the treetops." 1978. *Science Digest*, Oct. p. 26.

Engines of the Body

Job 10:10-11
"Hast thou not poured me out as milk, and curdled me like cheese? Thou hast clothed me with skin and flesh, and hast fenced me with bones and sinews."

The automatic transmission in a vehicle changes gears in response to changing power demands and driving conditions. Likewise, the muscles in your body change gears in a sense, in response to your changing needs for them.

When you use a muscle group, the muscles begin to need more fuel and more oxygen, both brought to them by your blood. When you use a set of muscles, lots of new blood vessels will actually grow into the tissue to meet the increasing demand. This is one way in which exercise lowers blood pressure – by increasing the "tubing" through which it passes. If you often overuse a group of muscles, additional connective tissue will grow through the muscles, strengthening them.

The "automatic transmission" in your muscles engages when you use the same group of muscles for a continuing task like rowing a boat. Before long, the normal fuel freely available in your blood is used up. At that point your body mobilizes stored fat, which allows the muscles to continue without exhaustion. Muscles that are often used to this point actually become more efficient at burning fat reserves. In even more strenuous exercise, a third "gear" kicks in, changing the way muscles burn fuel, so there is less waste for the circulatory system to deal with.

As scientists learn more about our incredibly complex and intelligently designed bodies, we find even more reasons to bear witness to our wonderful and wise Creator.

Prayer: Dear Lord, I thank You that You have so carefully and lovingly designed and built my body. Help me to take good care of it and glorify You with it. Amen.

Ref: Bodanis, David. 1985. "What to look for in home fitness machines." *Science Digest*, Apr. p. 42.

Whales Write Songs

Psalm 148:7
"Praise the LORD from the earth, ye dragons, and all deeps:"

Most of us have heard recordings of whales singing. Scientists have not yet learned to understand the whale's language, but they have identified individual verses in whale song. Now they are more convinced than ever that whale song is a sophisticated language.

Whale song is made up of complex and stylized compositions, many of them longer than the longest of human symphonies. A single whale song can last up to 22 hours! Scientists have learned that humpback whale songs change a lot from year to year. Yet each humpback in an entire ocean will always sing the same song as the others. Scientists wondered how whales could keep the verses straight since the song changes so often.

Researchers have concluded that whales, like human beings, use rhyme to help remember their songs. Biologists studying whale song report that they have identified the same sub phrases turning up in the same positions in verses that are next to each other. Whales apparently compose their long songs with rhymes, making them possibly the greatest poets on earth!

Evolutionists tell us that language is such an advanced development that even the first human beings could do no more than grunt and growl. And yet the whale not only has a sophisticated language but also creates poetry. Christians can comfortably put evolution aside to realize that the Bible is right when it says that the whole creation praises God. And whales make it very clear that the Creator Himself is the author of language – not some caveman.

Prayer: *Dear Lord, one of Your greatest gifts to us is language because through it we learn of Your love for us in sending Your son, Jesus Christ, to save us from sin, death and the devil. With that in mind, help me to be a good steward of the gift of language in praise and witness to You. Amen.*

Ref: *Science Frontiers*, July-Aug. 1989. p. 2. *Newsweek*, Mar. 20. 1989. p. 63.

Smart Heart

Hebrews 3:10
"Wherefore I was grieved with that generation, and said, They do alway err in their heart; and they have not known my ways."

People sometimes wonder why, with all the marvels science can design and build, it cannot build a good replacement for the heart. As medical science learns more about the heart, it becomes clear that replacing this complex mass of muscles is more difficult than might be imagined.

Your heart is about the size of your fist, really surprisingly small for the job it does. The heart is a pump with valves, tubing and an electrical system. Its four chambers are carefully designed to make sure that used blood goes to your lungs for oxygen and that blood from your lungs takes that oxygen to the rest of your body. In order to do this, your heart must pump 2,000 gallons of blood through more than 60,000 miles of soft tubing every day.

But the heart is much more than a pump. Researchers describe it as "an incredibly intelligent organ." The heart is smarter than we are. When you need more circulation, it senses this and speeds its pumping action. It doesn't matter whether the need for more pumping action comes from increased exercise, or something you thought about that might require more heart action, the heart responds. Even when transplanted, the heart is able to respond to its new environment.

Like everything else God has created, the heart is not a simple organ, limited to only one job. The heart is custom designed to do many jobs well. And to this day doctors cannot tell you why one apparently healthy heart stops beating and another does not. Life itself is more than an organ; it is a gift from our Creator to each of us.

Prayer: Dear Father in heaven, I thank You for the gift of life. Take my life and let it be consecrated to You. In Jesus' Name. Amen.

Ref: Huyghe, Patrick. 1985. "Your heart: a survival guide." *Science Digest*, Apr. p. 31.

New Technique Supports Creation

Psalm 119:160
"Thy word is true from the beginning: and every one of thy righteous judgments endureth for ever."

Some time ago, a news article described how scientists had studied the genetic material of a magnolia leaf that supposedly died 20 million years ago.

It might seem amazing that scientists would be able to duplicate the genetic material of a fossil, but the technique does seem to work. However, we reject the 20-million-year date in favor of a biblical history of life on Earth. In fact, what scientists found out about this leaf supports the biblical history.

The genetic information in living things contains the instructions for making that living thing. The color of your hair, eyes, skin, your height and all of your other features are contained in your genetic information. Scientists are not able to read that information in all its detail, but they can tell the difference between one kind of creature and another. They can also compare similarities and differences in great detail. But when they compared this supposedly 20-million-year old magnolia leaf to modern magnolia leaves, they discovered that even in its genetic detail it was virtually identical to the modern magnolia. Scientists freely admit their surprise. No evolution had taken place!

As science increases its ability to study the genetics of ancient creatures, it will continue to fail to find evidence of evolution. That's because all creatures were originally created in finished form, according to their kinds, just as the Bible says.

Prayer: Dear Father, Your Word is true! I ask that its truth would continue to challenge those who seek to explain the world without You. And I ask that its truth would become even more influential among Your people. In Jesus' Name. Amen.

Ref: Robinson, Jack. 1990. "Ancient leaf DNA copied at UCR lab." *The Press-Enterprise* (Riverside County, CA), Apr. 12. p. 1.

The Very Hairs of Your Head

Matthew 10:30
"But the very hairs of your head are all numbered."

In showing how the Creator is intimately involved with His creation, Jesus said that the very hairs of each of our heads are numbered by Him. No detail is too small to escape His attention; no change misses His careful and loving eye.

The hair that you see is nothing more than dead protein produced by the cells of the hair follicle anchored inside the layers of the skin. The total number of hair follicles in an adult is about 5 million, with only about 100,000 on the scalp. Each hair grows out of one follicle for about three to five years. Then the hair is shed and the follicle rests for about three months before starting to grow hair again.

So you see, even once you know how many hairs are on your head, it's no easy job to keep track of their changing number. The average scalp hair grows about one inch every two or three months. This means that each day your head is growing the equivalent of one 100-foot-long hair – that's nearly seven miles per year!

Yes, it's true; the Creator cares so much about you that He knows from moment to moment how many hairs are on your head. He has not created our world and then left us to drift through space and life all alone. He has even expressed His love and purpose for you in the words of the Bible. Take a look today at what Scripture says about His Son, who loved you so much that He even gave His life for you!

Prayer: I confess, dear Lord, that I cannot understand how You can keep track of all the details of the creation as Your Word says You do. But You are God and I am only human. Don't let my weakness limit or weaken my faith in your day-to-day involvement in my life. In Jesus' Name. Amen.

Do the Bible and Science Mix?

John 3:12
"If I have told you earthly things, and ye believe not, how shall ye believe, if I tell you of heavenly things."

Scientific principles learned in the Bible have led to countless scientific discoveries and saved millions of lives. It's true. Without the Bible, we would never have the blessings of modern science.

Isaac Newton became one of the greatest scientists in history because he learned from the Bible to look for intelligence and order in the handiwork of the Creator. Louis Pasteur knew from the Bible that life could not come from non-life. After all, he said, the Bible teaches that God is the Creator and author of life. Pasteur's scientific work laid the foundations for modern medicine and new techniques for storing food – both of which have saved millions of lives.

In the 19th century, Matthew Maury, the father of the science of oceanography, read in Psalm 8:8 that there are paths in the sea. Taking God at His word, Maury went on to discover the great sea currents that span the globe and nourish most of the oceans' life. He wrote: "The Bible, they say, was not written for scientific purposes, and therefore is no authority in matters of science. I beg pardon! The Bible is authority for everything it touches. The agents concerned in the physical economy of our planets are ministers of Him who made it and the Bible."

God has given us the Bible to make us wise unto salvation. But to paraphrase Jesus' words to Nicodemus, if the Bible tells us about earthly things and we don't believe it, how shall we be able to believe the Bible when it tells us about heavenly things?

Prayer: Lord, we believe; help our unbelief. Fill us with a new appreciation for your Word so that we may be instructed by You in all truth. In Jesus' Name. Amen.

Who Is God?

II Timothy 3:14-15
"But continue thou in the things which thou hast learned and hast been assured of, knowing of whom thou hast learned them; And that from a child thou hast known the holy scriptures, which are able to make thee wise unto salvation through faith which is in Christ Jesus."

Did you know that the Bible never tries to convince the reader that there is a God? As surprising as that sounds, it's absolutely true. The Bible's very first words begin by identifying God – but nowhere does the Bible try to prove there is a God.

The first verse of Genesis reads, "In the beginning God created the heavens and the earth." Here we learn that the God of the Bible is our Creator. We also see here, although less clearly until we look at the next two verses, the Prime Actor in creation – the Father.

In the second half of verse 2 we read, "And the Spirit of God was hovering over the surface of the waters." Now it is clear that the Trinity is being presented. Here is the Holy Spirit, moving over the as yet unformed Earth, anticipating the people yet to be made who would serve as His temples.

Verse 3 opens with, "And God said…" That simple phrase introduces the very heart of Scripture, the Word of God Himself. This is the very Word of God who would come and take upon Himself our earthly form in order to accomplish our salvation.

So even here in Genesis, we have the beginning of God's revelation of the Person and work of the Son of God – our Savior. Truly all Scripture has been given to make us wise unto salvation.

Prayer: *Dear Heavenly Father, without the revelation of Your love for us in Christ, I would consider myself lost and without hope or I might build my hope on false pride. So Your Word is a blessing to me in more ways than I can count. Thank You. In Jesus' Name. Amen.*

The Days in Genesis

Genesis 1:5
"And God called the light Day, and the darkness he called Night. And the evening and the morning were the first day."

Silently a huge, powerful form slides through the deep, cold, dark depths of the sea. The men aboard the nuclear submarine have seen neither sun nor daylight for months, yet each one knows what day it is. The men know what day and what time it is even without seeing daylight, because the sun's movement – like a clock – only measures time; it doesn't create it.

God doesn't need the sun to measure time either. When He tells us in Genesis 1 that He created everything in six days and rested on the seventh, we know these are days like ours, even though the sun was not created until the fourth day. Some people wonder whether the days of Genesis 1 could be figurative days. Well, the best interpreter of Scripture is Scripture itself. What does it say?

The word translated "day" in Genesis 1 is the Hebrew word *yom*. Whenever that word is used anywhere else in the Old Testament with a number – like 10 *yoms* – it *always* means a 24-hour day. And whenever *yom* is used anywhere else in the Old Testament with the phrase "evening and morning," it *always* means a 24-hour day.

Going back to Genesis 1, we see that the Holy Spirit has made sure that *both* of these rules are in force to assure us that the Genesis days are like ours!

Prayer: I thank You, Lord, that Your Word is clear and true. Use it to correct my understanding as well as my life and do not allow my own pride to make me deaf or blind to Your Word. For Jesus' sake. Amen.

Ref: Bartz, Paul A. 1988. "Days in Genesis one and the week." *Bible-Science Newsletter*, Aug. p. 10.

God Shows Us the Earth from Space

Job 26:7
"He stretcheth out the north over the empty place, and hangeth the earth upon nothing."

The Earth floats in space, attached to nothing, surrounded by a thin layer of air. What science has only just learned, the Bible has taught for thousands of years! Yes, while other ancients pictured the world as flat or resting upon giant turtles or some other animal, God told the Jews in Job 26:7 that He "hangs the earth on nothing."

In Genesis 1:6 we read that God created a firmament. In recent times some have said that this word proves the Bible is based on ancient myths. New discoveries, however, are challenging these doubts about the Bible.

The word translated "firmament" from the Hebrew *ragia* in these verses comes from a Hebrew root word that refers to the process of making a statue. In making a statue, the ancient artisan would take a soft metal – like gold – and begin to carefully pound thin sheets of it onto a wooden form of the statue until the wood was completely covered by a thin, form-fitted layer of gold.

The use of this word puzzled many people until the Earth was first viewed from space. Then we saw it – the Earth suspended on nothing in space, surround by a thin, form-fitted layer – our atmosphere! So the Bible tells the truth in all the subjects it mentions. But no matter how long science studies, it cannot learn about God's love to us in Jesus Christ. This is revealed to us only by the Bible!

Prayer: Dear Father in heaven, there is no place that man can go that You have not already been there; there is no knowledge man can have that You don't already know. Grant Your Holy Spirit and wisdom to those of us who are called by Your Son's Name, so that we may not be misled in these confusingly challenging times. For Jesus' sake. Amen.

In the Midst of the Waters

Genesis 1:6
"And God said, Let there be a firmament in the midst of the waters, and let it divide the waters from the waters."

What was the Earth like before the Flood? Bible-believing scientists are giving us some surprising answers about the incredibly beautiful Earth God originally created.

In Genesis 1:6 we read that God divided the waters, leaving waters above and below the firmament. The firmament spoken of here is our atmosphere. We can easily understand that the waters below the firmament are the oceans. But what are the waters above the firmament?

The most commonly accepted theory offered by Bible-believing scientists is that the waters above the firmament may have been a water vapor canopy. A water vapor canopy above most of the atmosphere would have had the same effect as the roof of a greenhouse does today. The Earth would have been a warm, tropical place, with very even temperatures from pole to pole. Under such conditions there would be no storms nor winter as we know it. This theory, say creation scientists, would explain why we find evidences of tropical plants and animals even in the far north and on the Antarctic continent.

Creation scientists have suggested that Genesis 7:11 may be referring to the collapse of this canopy when it says that the windows of heaven "opened up." Yes, the Bible offers us a believable history of major events that can be accounted for in only thousands of years instead of millions!

Prayer: Dear Lord, I thank You that Your Word is trustworthy and true. Let its truth be evident to all so that more may join their voices in glorifying You! Amen.

Ref: Bixler, R. Russell. 1986. "Does the Bible speak of a vapor canopy?" *Bible-Science Newsletter*, Nov. p. 1.

After Their Kinds

Genesis 1:12
"And the earth brought forth grass, and herb yielding seed after his kind, and the tree yielding fruit, whose seed was in itself, after his kind: and God saw that it was good."

How wonderful! Your dog has just had puppies! But do you now have to sort through the litter and make sure there are no baby giraffes or kangaroos?

In God's account of creation in Genesis 1, we repeatedly read that both plants and animals were created to reproduce "after their kind." Genesis 1, in speaking about the creation of plants, repeats three times in just two verses that they are to reproduce "after their kind." We see the same phrase repeated later in the chapter when animals are created. This is not just empty repetition. God is stressing a fundamental principle that all things reproduce "after their kind." Mother dogs have puppies and mother cats have kittens. You can count on it.

Why does God stress this principle? Even before creation, God knew that humans would eventually sin and then seek to hide their responsibility by trying to explain things without a Creator. God knew that this idea of evolution would capture the faith of millions over the history of the world.

God stresses what our experience shows so that He might be harder for us to hide from. All things do reproduce after their kind. And despite evolutionists' strong faith in evolution, they cannot offer one established scientific fact to explain how one kind of creature might eventually make a completely different kind!

Prayer: I thank You, Lord, that You have made it hard for men to deny You. Yet men still deny You, seeking explanations and excuses outside of Your Word. I know that I, too, can do this, for I am both saint and sinner. I ask that You would correct me when I seek outside of Your Word what is already so richly provided to me in Your Word. Amen.

The Sun, Moon and Stars

Psalm 8:3-4
"When I consider thy heavens, the work of thy fingers, the moon and the stars, which thou hast ordained; What is man, that thou art mindful of him? and the son of man, that thou visitest him?"

What is the most awesome show of God's power? It may not be what you think.

In Psalm 8:3-4, the psalmist is led to exclaim, "When I consider Your heavens, the work of Your fingers, the moon and stars which you have ordained, what is man that You are mindful of him…?" If the night sky is a glory we can only stare at in awe, our telescopes and space probes have shown us that we can see very little of its true glory.

Consider our sun. Less than 0.10 percent of all the sun's energy falls on the Earth. Yet if just that small fraction of power could be harnessed, we would never have any energy shortage. But we have learned that our sun is only an average-sized star in our galaxy of over 1 billion stars! We have no way to even measure that kind of energy! Even more awesome is that our galaxy is only one of more than a million galaxies! What is a billion times immeasurable energy? And God made and energized it all in just one day!

As hard as all of this is to understand, what is even harder for us to comprehend about God's working is that all of this was created through the power of God's Word – that same Word that became flesh and dwelt among us! Truly, His love for us is beyond our comprehension!

Prayer: Dear Father, though I cannot comprehend it, I thank You for Your Love for me which moved You to send Your only Son for my redemption. Help me to better understand such love as You have and make me better able to show it to others. In Jesus' Name. Amen.

For the Sheer Joy of Variety!

Isaiah 45:9
"Woe unto him that striveth with his Maker! Let the potsherd strive with the potsherds of the earth. Shall the clay say to him fashioneth it, What makest thou? or thy work, He hath no hands?"

Did you ever try to plan all the details of a simple project? How many plans do you think the Lord had to make when He created living things? A billion? A billion times a billion?

We all know that it takes time to plan the most simple project. Did you ever think about the planning God had to do when He created all the different kinds of living things? Our word "species" today includes many creatures that the Bible counts as being the same "kind" – as when God created the kinds. But God did design the genetic information that allowed the kinds to produce these variations.

Yes, God's act of creating living things was much more than just wishing. Just think that there are more than 20,000 different species of bees – some with very complex societies – and their own languages! The figures and the beauty of it all makes one wonder at God. Why are there 4,500 different species of sponges? Why are some creatures – never seen by humans until this century – so eerily beautiful? For that matter, why have so many different kinds of beautiful flowers?

The variety in the creation reflects some of the joy of creation that God felt, and shows us the incredible unbridled creativity of our wonderful God. The fact that there is only one species of human – all related – confirms the human history related in the Bible.

Prayer: Dear heavenly Father, I know that I shall never have Your ability to plan and carry out those plans. I confess that too often I waste the time and energy You give me because I don't even bother to use the abilities You have given me. Forgive me for Jesus' sake, and for His sake help me to be more like You in this. Amen.

Humanity – Past, Present and Future

Genesis 1:26
"And God said, Let us make man in our image, after our likeness: and let them have dominion over the fish of the sea, and over the fowl of the air, and over the cattle, and over all the earth, and over every creeping thing that creepeth upon the earth."

A straightforward reading of Genesis offers a very different story of humanity than does modern evolutionary science. Does the rest of the Bible contradict evolution too? Can Genesis and evolution be harmonized?

According to evolution, humans are the result of millions of years of life, struggle and death. Today, we are little more than a subchapter in that long story of endless struggle and death. Can this be reconciled with the Bible? Not if we let the Bible interpret itself. First, the Bible allows for only one day of history for animals before humans appeared on the scene. Second, humans were created not from some other creature but were handmade by God, in His image.

The most important difference between evolution's story and the Bible's story of humanity is the role death plays. According to evolution, death was already part of nature long before humans ever showed up. According to the Bible – for example, in I Corinthians 15:21 – death came upon creation because of the first man, Adam's, sin. This is why it was necessary for another man, Jesus Christ, to do away with death.

For the Christian, the most objectionable part of evolution is that it separates sin and death from each other. This makes Christ's death and resurrection for us completely redundant, since death has nothing to do with sin! There can be no harmony between this and the Gospel!

Prayer: Dear Father, You specially created human beings because You desired to have a personal relationship with each one of us. As I consider my spiritual relationship with You, help me to remember that Your Son, Jesus Christ, died so that I might, through Him, have the forgiveness of sins. In His Name. Amen.

Does Genesis 2 Contradict Genesis 1?

John 10:35b
"...and the scripture cannot be broken;"

As we read Genesis 2, we might get the idea that humans were created before the animals and even before plants. Since this would seem to be a clear contradiction of Genesis chapter 1, some have said that the creation account does not intend to offer literal history. Is this really the case?

The reason for these apparent differences becomes a little more clear to us when we realize that in some languages there is no apparent contradiction. That tells us that the reason for these apparent contradictions has something to do with how English works, not with what the original says.

That is indeed the case. Our English language has time built into our verbs. Past, present and future – we all learned these in school. But Hebrew verbs – the language in which these verses were originally written – do not have time built into them. So this problem always shows up when we try to express these thoughts in a language that has time in its verbs. Genesis 1 is very careful to express time relationships; each day is numbered. Genesis 2 is interested only in focusing on the details of the human story. The other details, covered in sequence in Genesis 1, are mentioned only as they serve to add to the story of humanity.

So there is no contradiction between Genesis 1 and Genesis 2. It only appears so in English because we have no such thing as a verb that does not express time. God's Word stands as completely trustworthy!

Prayer: Lord, I marvel and give thanks to You for the careful accuracy of Your Word. Help me to apply myself to a more complete study of Your Word and grant me Your Holy Spirit so that I may understand and believe what I learn. Amen.

Ref: Niessen, R., B. Northrup, and D. Watson. 1988. *Genesis Stands*. Minneapolis, MN: Bible-Science Association, Inc.

Are Dinosaurs a Giant Mystery?

Genesis 6:20
"Of fowls after their kind, of cattle after their kind, of every creeping thing of the earth after his kind, two of every sort shall come unto thee, to keep them alive."

Can you picture Noah trying to get some of those giant dinosaurs on the ark to save them from the Flood? There's no denying that dinosaurs really lived. The land dinosaurs were likely created on the sixth day with the other great land creatures, while the marine dinosaurs were created on the fifth day with the other denizens of the deep. So what happened to them? Today there seem to be more theories about why dinosaurs disappeared than there were dinosaurs to disappear.

One of the most popular theories is that a huge rock from space, perhaps miles across, struck the Earth. The fires, earthquakes and storms that followed so polluted the atmosphere that plants and then dinosaurs died. Evidence for this theory is lacking; every dinosaur bone found today is in rock that bears evidence of having been buried rapidly by a great amount of water. That sounds like a Flood to creation scientists.

The dinosaurs are acknowledged to be a type of reptile and are known to have laid eggs. Many hundreds of fossilized dinosaur eggs have been found. All this means that even the largest of the dinosaurs began life as little fellows. Noah would have had no problem looking after them on the ark – they were likely no larger than the family dog. And how did the dinosaurs become extinct? Probably the same way the Dodo and the carrier pigeon became extinct – at the hand of humans.

Because the world's climate after the Flood was cooler and wetter, dinosaurs who survived the Flood did not prosper, and most – if not all – of them have become extinct. Dinosaurs are yet another –to the incredible wisdom, power and diversity of God's wonderful creativity.

Prayer: Dear Heavenly Father, not only is Your power shown in the great dinosaurs You created, but in the Flood as a judgment on man's sin. Do not abandon this age of Your people to their sin; rather, use those who are called by Christ's Name to hold Him up as the Savior of the world. Amen.

The Work of a Superior Intelligence

Romans 1:20
"For the invisible things of him from the creation of the world are clearly seen, being understood by the things that are made, even his eternal power and Godhead; so that they are without excuse:"

If we discovered the work of an intelligence so superior to our own that we could not understand it, would we even recognize it for what it was?

If we were faced with the work of such an intelligence, what would we think of it? Consider an African bushman examining a simple calculator. All he would see was a flat box that had no apparent purpose. If he opened it up, all he would find would be a seemingly purposeless jumble of wires. He might even decide that the object was something natural – perhaps something that grew on a tree – rather than the careful design and work of another human.

When we are faced with the intelligence of God in the creation, it is often easy for us, like the bushman, to see things we cannot understand as just happening. This is really what the origins debate is all about. But nothing just happens. While the leg of an ant appears to be a simple affair – quite natural to us – the Soviets learned differently. Their lunar vehicle, the Lunakod, was designed to walk on the moon with articulated legs designed after the ant. Designing the vehicle inspired a new respect among Soviet scientists for the leg of the lowly ant – for it was difficult to design and took many years to produce.

In Romans 1 God says that His power and majesty are evident from what He has made in the creation. We are learning that nothing just happens; even the simplest things require great work and effort to design and build.

Prayer: Dear Lord, truly those who have decided that the beautiful creation in which we live could have been made by nothing have fallen under a delusion. But I know that this delusion of sin is one which none of us is immune to, which is why I and all men must depend completely on the forgiveness of sins You won for us on the cross. Amen.

The World's Smallest Computer

Exodus 15:11
"Who is like unto thee, O LORD, among the gods? who is like thee, glorious in holiness, fearful in praises, doing wonders?"

The cells of all living things are filled with highly detailed information. The incredible system that is used for storing this information makes our best computer information systems look like a child's simple slate board!

Believe it or not, it takes more information to make a complete human being than can be stored in the average city library. And all of this information is stored in less than a few thousand-millionths of a gram of material within each of our cells. You have literally billions of copies of it! Our most sophisticated computer system cannot even come close to being able to store information like that.

It is estimated that one billion species of plants and animals have existed since creation. The information storage system within the cell is so compact that all of the information to make every one of those one billion species could be held in a teaspoon – with room left over to hold all the books ever written if they, too, were stored with the same efficient system. Even more amazing is that this information storage system is not only able to store information like a computer but, unlike a computer, it can also copy itself and correct errors if any creep into the information!

How could anyone believe that this brilliant information storage, copying and correction system could have come about by chance? To believe that this could be created by no one is certainly an act of faith. But it is clearly the wrong faith! And that highlights what the origins debate is really all about.

Prayer: Father, I believe in Your forgiving grace to me in Jesus Christ. For this reason, help me to abandon all trust in myself. For Jesus' sake. Amen.

Can There Be Life Without God?

Job 10:9
"Remember, I beseech thee, that thou hast made me as the clay; and wilt thou bring me into dust again?"

Did life form all by itself, without a Creator, in some warm little pool of water or a clod of mud many billions of years ago? Some scientists would like us to think so. Over the years, newspaper headlines have declared that life has been created in the laboratory. We are told that life is no mystery and that the chemicals that cause life are common.

Such claims have a lot of "hype" and little science to them. There are some things you should know about these experiments. First of all, scientists don't end up with life. They do end up with some simple chemicals that are like those found in our cells. But the chemicals they make, called amino acids, are a mixture that is quite worthless to form life.

In order to get these amino acids, which are supposed to show that life could form by itself, researchers use a special combination of gasses that they believe formed the Earth's atmosphere millions of years ago. They must also carefully control the kind of light that enters the experiment and protect in a special container anything that is produced. It is obvious that a lifeless Earth did not offer all these advantages.

It seems silly to invest millions of dollars worth of equipment, dozens of years of specialized training, more years of research and then say that we are proving that no intelligence or design was needed to create life! What these experiments do show is that life cannot happen by itself – it took a Creator with much more power and wisdom than we have to create life!

Prayer: In Your wisdom, dear Lord, You have created life in such a way that even those who are most intent on denying You cannot comfortably do so. Use my voice and life to help them see that this is because You have loved them and would be their Savior too. Amen.

There Is No Simple Life

Job 37:8
"Then the beasts go into dens, and remain in their places."

We often hear people speak of "simple life forms" or the "simple cell." Is there really such a thing as "simple life"? Although we often speak of the "simple cell," there really is no such thing as a simple form of life. Even the single-celled creature is a highly complex organism.

Let's imagine building a model of the cell, using atoms, each the size of a tennis ball. The typical cell has about 10 million atoms. In our model, the cell would be over ten miles in diameter! Now let's look inside the cell and see whether it is simple or not. First of all we would see lots of activity going on almost everywhere. This is because this single cell has to carry out all the same jobs that your body uses many different organs to accomplish.

Inside the cell we would see structures that digest food and others that carry off the waste from food. Other structures would be busy carrying off the waste produced by the cell itself. The cell must also contain everything it needs to reproduce. And remember, while we have many complex and different organs to do all this, the cell must to it all without those organs.

Humans have never even come close to creating anything like this. Surely the cell is not simple! Science cannot explain the cell by any natural laws it knows. Scripture is correct when it tells us that the origin of the cell – and everything else – is the superior power and wisdom of God.

Prayer: Dear Heavenly Father, You are defamed when the world talked of even one living cell as "simple." If someone spoke so of my best friend, I would speak in his defense. So let me boldly speak in defense of Your truth, my only Heavenly Father. In Jesus' Name. Amen.

Should the Sun Spin Faster?

Psalm 74:16
"The day is thine, the night also is thine: thou hast prepared the light and the sun."

According to evolution, some explanation for the sun and planets must be found that does not include God as the Creator. Today, evolutionists generally believe that the sun and the other planets of the solar system are related to each other.

One of the most popular theories is that the sun and the planets of the solar system each formed when a single cloud of dust and gases were drawn together by gravity. As this material was drawn together, it began spinning faster and faster, like a giant skater pulling his arms toward his body. According to this theory, the amount of this angular momentum – which we will simply call "spin" – among the planets must equal the amount of "spin" the sun has.

As we have gradually learned more about the solar system, we have been able to compute this energy more accurately. Figuring in the mass of the planets, their orbits and the mass and rotation of the sun, scientists have concluded that 98 percent of the solar system's "spin" is in the planets. Only two percent, not the 50 percent expected, is in the sun.

So modern science has confirmed – probably as much as it can in this case – that the Earth was created independently from the sun. Of course, we know that Scripture clearly says that the Earth was created three days before the sun. Here, a scientific mystery for evolution is easily explained when we turn to the Bible!

Prayer: Dear Heavenly Father, Creator of all stars, I give thanks to You that You have shown Your great power in the stars, but that You came to us not to bring fear but Your undeserved love through salvation in Your Son, Jesus Christ. In His Name. Amen.

We See by Faith

Hebrews 11:1
"Now faith is the substance of things hoped for, the evidence of things not seen."

As Christians, we believe by faith. But did you know that even evolutionary scientists believe by their own kind of faith? Scientists early in the previous century, like scientists today, had no evidence that humans had evolved from ape-like creatures.

In 1922 in Nebraska, a tooth was discovered that was said to have belonged to a missing link between human and ape. But what did this creature look like? As is done today, paleontologists began to "reconstruct" "Nebraska man." They reconstructed what the jaw bone around the tooth might have look like, and then the bone touching those bones, and – well, you know how the song goes: "The head bone's connected to the neck bone; the neck bone's connected to the back bone…"

Before long they had constructed, from that one little tooth, not only what "Nebraska man" looked like, but also what his wife looked like. And they put this "proof" in museums and textbooks. Eventually they discovered more parts of the animal from which the tooth had come. It turned out to be the tooth of an extinct pig! But even this was wrong – in 1972, living herds of this very pig turned up in Paraguay!

Anyone, including scientists, can make mistakes. But what this true story shows is how, by rejecting God's account of creation, a pig could be made into a human. You see, even the evolutionist's belief is not based on scientific fact but on his own faith in nature rather than in the Creator. We Christians should not be ashamed to admit that we believe by faith, because our faith is built on the solid statements of the uncontradicted Scriptures!

Prayer: Dear Father, I thank You that You have allowed me to hear Your Word and that You have given me faith in Your promises. Teach me, through Your Word, so that I am better able to identify false religious beliefs and carry Your witness. In Jesus' Name. Amen.

Evolutionists Love "Lucy"

Genesis 2:7
"And the LORD God formed man of the dust of the ground, and breathed into his nostrils the breath of life; and man became a living soul."

"Lucy" is said by its discoverer, Dr. Donald Johanson, to be the best and latest proof that humans evolved from some ape-like creature. But Lucy is really a monument to *human* imagination!

In 1978, Dr. Johanson announced his discovery of Lucy to the world at the Nobel Symposium on Early Man. Every newspaper quickly picked up the story. Lucy was supposed to be the earliest human ancestor. Dated by evolutionists at about 3 million years old, Lucy was said to be an almost complete skeleton of a creature that walked upright like humans. However, there are some facts you should know.

Lucy is really a demonstration of how creative people can be – sort of tinker-toy science. For example, her knee comes from a location over 200 feet lower in the rocks and over a mile away from one of her leg bones! Lucy has been assembled from materials found in two different locations separated by several miles. Her bones were sifted out of plots of ground at these two sites, each of which was almost the size of a football field!

Strangely enough, all of Lucy's bones from one of the locations provide her human-like characteristics, while all of the bones gathered from the other location provide her ape-like characteristics. That sounds at least a little suspicious!

Prayer: Dear Heavenly Father, You are our true Father. Today I especially ask that those Christians who are being enticed into believing evolutionary stories be made to understand that the facts name You, and not some ape, as our true Father. In Jesus' Name. Amen.

Ref: Willis, Tom. 1987. "'Lucy' goes to college." *CSA News*, Feb. p. 2.

Morality Is for Humans

Leviticus 11:44a
"For I am the LORD your God: ye shall therefore sanctify yourselves, and ye shall be holy; for I am holy:"

Is right and wrong really a matter of doing what *you* think is best? What does morality have to do with origins?

We often hear today that right and wrong is really a matter of doing what *you* think is right and wrong in any given situation. This idea that right and wrong depends on what you think and the situation comes from the view that humans are nothing more than evolved animals. According to this idea, we have no one to be responsible to but ourselves. We must explore and discover for ourselves what is right and what is wrong.

But according to the Bible's teaching about creation, we were created by an all-wise and powerful God. He designed and made us, and He knows which activities are destructive to us and which are good for us. Beyond that, being our Creator, He owns us – lock, stock and barrel! He has all rights over us, including the right to hold us to account for our decisions and actions. His commandments and rules for living are given to us for another reason; He knows that the activities and attitudes they encourage are healthy for us, while breaking His laws hurts us. He doesn't want to see us hurt.

So God's commandments are absolutes that we are responsible to keep. Unfortunately, we have all broken these. Fortunately, He knows our condition and still loved us enough to send His only Son to rescue us from our sin. Creation shows us what our sin and our salvation are based upon and what they really mean to creatures of God.

Prayer: Dear Father in heaven, although Your law reflects Your will for us so that we do not destroy ourselves, at the same time it accuses us as sinners, for none of us can keep Your law perfectly. For this reason I thank You that You have sent Your Son, Jesus Christ, so that through Him I am forgiven and restored to You. Amen.

Creation and the Gospel

John 1:3
"All things were made by him; and without him was not any thing made that was made."

What does the Bible's teaching on creation have to do with the Gospel of Jesus Christ? Are these just two separate and unrelated biblical teachings? Christians often ask us, "Shouldn't we just worry about the basics? Why worry about a teaching like creation?" Indeed, we should be concerned with basics.

But what happens to Christ's work of salvation if creation is not true? If evolution is true, then death came into the world long before the first humans and their sin. If death is not a result of sin, why did Christ have to receive the penalty of sin – death on a cross. So in challenging human origins, evolution challenges the origin of sin and its effect on humanity. And in challenging this, evolution challenges the very reason Christ came to Earth! He becomes just a misguided being on the wrong planet! The Bible tells us that all Scripture was given to make us wise unto salvation. That includes Genesis. The first chapter of St. John's Gospel tells us that all things were created through the Word – the Word who became flesh.

We see that Word in action in Genesis 1 when we read, "And God said…" That Word who made us and everything is the very same Word who came and purchased our salvation.

So Genesis, beginning with the first chapter, is actually the beginning of God's revelation to us of the Person and work of the Son of God – our Savior Jesus Christ. If we reject God's revelation about our Savior in Genesis 1, we have only part of a Savior – and only part of Christ is no Christ at all!

> **Prayer: Dear Lord Jesus Christ, many in our world today attack You by denying Your work of creation, hoping through that to deny Your wonderful saving work for sinful mankind. This even happens in the Church. Give Your people eyes to clearly see the truth and not be led into error and left with a powerless word of man in place of the Gospel. Amen.**

The Amazing Woodpecker

Deuteronomy 4:28
"And there ye shall serve gods, the work of men's hands, wood and stone, which neither see, nor hear, nor eat, nor smell."

The woodpecker's tongue can stretch from three to five times its normal length in order to fish bugs out of trees. You would think that his tongue would have to be rooted in his tail to do that neat trick! This beautiful creation in which we live is not only filled with testimonies to the wisdom of God. It is also filled with special designs that deny the possibility that life in all its complexity developed entirely by chance.

Consider the woodpecker. Incredibly, the woodpecker's tongue is rooted in his right nostril. Exiting the back of the nostril, his tongue spits into two parts, wrapping around his head between his skull and the skin, passing on either side of the neck bones, and then coming up through his lower jaw or beak. This gives the woodpecker a long enough tongue to stretch it out far enough to do an effective job of pest control on bug-infested trees!

Now how could this happen by blind evolutionary chance? Even evolutionists admit that it's silly to suggest that the woodpecker's tongue gradually got longer over thousands of year and began to grow under his skin. As one evolutionary scientist said about the woodpecker's tongue, "There are certain anatomical features which just cannot be explained by gradual mutations over millions of years. Just between you and me, I have to get God into the act too sometimes."

Why wait to call on God as a last resort? Let's begin with our wonderful Creator!

Prayer: Dear Heavenly Father, through the instruction of Your Word and the guidance of Your Spirit, help me to be different from those around me who think that the creation itself made them. In Jesus' Name. Amen.

Ref: Sunderland, Luther D. 1975. "Miraculous design features in woodpecker." *Bible-Science Newsletter*, v. 13, n. 10, Oct. p. 4.

The Hi-Fi Cricket

Psalm 71:19
"Thy righteousness also, O God is very high, who hast done great things: O God, who is like unto thee?"

What has two well-designed horn-shaped speakers that are designed to amplify and project sound in multiple directions?

No, we're not talking about your friendly neighborhood teenager. We're talking about the male mole cricket. The male mole cricket has a dandy design feature that can only be explained as the result of thoughtful planning and design – and a good working knowledge of acoustics. Random evolutionary processes can't explain this one.

When he is ready to mate, the male mole cricket sticks a pair of horns out of his burrow. The shape of these stereo horns is basically smooth, with an even shape that actually amplifies his mating sound. His two-speaker system also does something else that is quite unusual for living things – it projects sound evenly in all directions. Because there are two speakers, the female mole cricket is easily able to find the very center of the sound – the male mole cricket's burrow!

When one realizes that it took humans until the 20th century to learn how to actually apply these principles in their stereo systems, it becomes harder than ever to believe that unguided mutations and mindless chemical reactions came up with this design. God invented the stereo hi-fi before we did!

Prayer: Dear Father in heaven, You have provided for all of the needs of all of the creatures You have made. Help me to be equally concerned about what You have made, knowing all the while that You care for all my needs too. In Jesus' Name. Amen.

Ref: "Design in lower animals." *Five Minutes with the Bible and Science* (Bible-Science Association, Inc., Minneapolis, MN), v. 5, n. 11, Nov. 1975. p. 2.

Clean the Blushing Fish

Job 36:22
"Behold, God exalteth by his power: who teacheth like him?"

Many types of animals help each other – an activity called "altruism" – even though some of these animals would normally eat each other. Different kinds of animals that help each other pose a serious problem for those who deny an all-wise Creator. Evolution has lots of reasons why natural enemies would not cooperate to help each other. But consider the example of the yellow-tailed goat fish.

The yellow-tailed goat fish is a mostly white fish that swims in small schools. They often cruise around reefs where the yellow French angel fish live. The angel fish hide from the goat fish who might eat them. However, the goat fish are often bothered by parasites that lodge in their scales and gills. When this happens, the goat fish swim to the reef in which the angel fish live and blush a bright rust red color.

When this happens, the angel fish knows that the goat fish has not come for lunch but to be cleaned. So the angel fish swims out and cleans the goat fish of his uncomfortable and unhealthy parasites! When the goat fish is clean he stops blushing and swims off to leave the angel fish in peace.

Is it possible that our wise Creator has provided for the needs of His creatures in a way that could provide these unexplained mysteries for those who want to explain things without Him? We think so. We think this is one of the ways in which His glory, power and wisdom are made evident to those who don't want to admit an intelligent Creator.

Prayer: Lord, I thank You that You leave those who are proud in the imaginations of their own hearts unsatisfied, so that they may continue to seek You, while satisfying those who look only to You. Help me always to seek You first in all things. Amen.

Fungus that Blows Its Top

Genesis 1:11
"And God said, Let the earth bring forth grass, the herb yielding seed, and the fruit tree yielding fruit after his kind, whose seed is in itself, upon the earth: and it was so."

God doesn't make cheap things! While we speak of "simple" forms of life, the more we learn about living things, the more clearly we see that there are no simple forms of life.

Evolution has long theorized that life began with simple forms that gradually, over millions of years, became more complex. Evolutionary scientists say that the "simple" forms of life still around today prove their claim. The problem is, there aren't any simple forms of life! Take the "lowly" fungus, for example. The cap-throwing fungus has a number of dandy designs that enable it to spread its reproductive spores. First of all, the cap-throwing fungus has a built-in clock. The fungus bends through the day in response to the sun's movement. It also throws its masses of spores out in order to spread them over the widest possible area.

The built-in clock of the cap-throwing fungus waits to blow its top until the fungus is turned at the best angle to produce the widest possible spread of its spores. The light-sensing system in the fungus releases the spores at about nine in the morning – aiming the spores at an area that is most likely to be open, so the spores can be spread even further by animals. But whether the spores land where animals pass or on a leaf, they are coated with a glue to aid in further dispersion.

The lowly fungus reminds us that there aren't any simple forms of life. Even this fungus has been given sophisticated ways to carry out its command to reproduce after its kind.

Prayer: Dear Father, there are none of us who are so simple or unimportant You have not enabled us to do Your will. Help me to always remember that the hindrance of sin is of my making, and to come to my Savior, Jesus Christ, for cleansing. In His Name. Amen.

Ref: Howe, George F. 1972. "The cap thrower fungus." *Creation Research Society Quarterly*, v. 9, n. 3, Dec. p. 172.

The Bat's Special Radar Design

Matthew 6:27-29
"Which of you by taking thought can add one cubit unto his stature? And why take ye thought for raiment? Consider the lilies of the field, how they grow; they toil not, neither do they spin: And yet I say unto you, That even Solomon in all his glory was not arrayed like one of these."

If you wake up when it's still dark, you know how overwhelming it can be to turn on the light. Your eyes may even hurt at first. If you've had this experience, you can understand one of the problems that had to be solved when radar was developed.

Basically, radar is made up of two parts. The transmitter sends out a powerful beam of radar waves. But the other part of the radar system is a very sensitive receiver that cannot stand the powerful outgoing signal. One of the major technical problems to be overcome in developing modern radar involved sending out this powerful signal without overwhelming the sensitive receiver. What scientists finally developed was a fast switch that turns the sensitive receiver off every time a radar pulse is sent out.

Bats, who have their own sonic radar, had this problem solved from the beginning. They have muscles in their ears that are the receivers for the echoes. These muscles close the ears for split seconds when the bats are sending out their high-pitched signals.

Without this feature, the bat's navigational system would be useless. How could a bat figure out that it needed this ability and then decide to grow the muscles and related tissue to do the job? When one decides to say that creatures, instead of the Creator, have made themselves, one can end up with some very silly conclusions!

Prayer: Dear Father, You have made all things well and with the good of the creation in mind. Help me to remember this when I tend to think of Your Word as separated from the realities of everyday life, thereby missing out on so many of the blessings You have prepared for me. In Jesus' Name. Amen.

Ref: Henson, O.W. 1971. *Journal of Physiology*, v. 180. p. 871.

The Giraffe's Wonder Net

Job 9:10
"Which doeth great things past finding out; yea, and wonders without number."

Did you ever stand up a little too quickly and get dizzy? That happens because by standing you may have temporarily lowered the blood pressure in your brain. Can you imagine what can happen when the giraffe swings his head from the ground to the treetop?

The giraffe's heart and the rest of his cardiovascular system is very different from ours—and from most other creatures. If it weren't different, there wouldn't be any giraffes! In order to get his blood all the way from his heart up that long neck to his brain, the giraffe's heart must produce twice as much blood pressure as would be expected in an animal his height.

But the giraffe's brain is very sensitive to high blood pressure. So giraffes have been given a special structure that's been called the wonder net, where the blood supply enters his brain. This "wonder net" keeps the blood pressure in the giraffe's brain just where it should be. Even if the giraffe drops his head quickly, say from nibbling a treetop to take a drink of water, the blood pressure in his head remains safe. The wonder net can quickly control such rapid changes. And to prevent used blood from draining back into his brain when he lowers his head, the giraffe has a special set of one-way check valves in his neck.

When we talk about our Creator, we need not be afraid that some people will think we are talking about worthless things. Truly the wonders the Creator has made are great and worthy to be told to people far and wide!

Prayer: Dear Lord Jesus Christ, give us the bold conviction we need to boldly tell others what You have done – from our creation to our salvation. In Jesus' Name. Amen.

Ref: "Sticking out his neck." *Creation Science Report* (Creation Science Research Center, San Diego, CA), July 1974.

The Harvesting Ant

Proverbs 6:6-8
"Go to the ant, thou sluggard; Consider her ways, and be wise: Which having no guide, overseer, or ruler, Provideth her meat in the summer, and gathereth her food in the harvest."

God cares about all of the creatures He has made – even ants. God has invented such incredible ways to provide for them that Bible skeptics often get caught in their own traps!

Those who don't believe the Bible is the Word of God have in the past pointed out an error, as they call it, in Proverbs 6:8. Proverbs 6:6-8 talks about the wisdom of the ant in gathering and storing food for the winter. Skeptics said this must come from some ancient myths about ants, because there are no ants known to gather and store food. In the last few years, however, their laughter has stopped. We now know of three different species of ants that gather and store food. And guess what? Two of these species are native to Palestine!

Evolutionists tell us that they are sure the first real human beings weren't even smart enough to gather and store grain and seeds. It is obvious that ants who gather and store grain and seeds are totally unexpected by evolution. But we know that we have a Creator who provides for all the needs of all His creatures – even ants.

The Bible shows itself to be true in all earthly things it tells us about. This is God's way of impressing upon us the accuracy of His Word, so that we might believe when it tells us about spiritual things like our Savior, Jesus Christ.

> *Prayer: I thank You, Lord, that You have not given us such wonderful brains and then asked us to ignore them and blindly believe. In Your wisdom, You give reasons to believe to those who believe and leave those who refuse deluded by their own wisdom so that they might seek You. Lord, I believe; help my unbelief. Amen.*

A Desert Traveler's Friend

Isaiah 41: 19-20
"I will plant in the wilderness the cedar, the shittah tree, and the myrtle, and the oil tree; I will set in the desert the fir tree, and the pine, and the box tree together: That they may see, and know, and consider, and understand together, that the hand of the LORD hath done this, and the Holy One of Israel hath created it."

What two things are most important to someone lost in the desert? Why, a compass and, of course, some water. If you are lost in the desert and find a compass barrel cactus, you have both in this amazing plant.

There are many true stories told about people who were lost in the desert and dying of thirst until they found the compass barrel cactus. Some of them expected to find an actual barrel of water inside the cactus when they cut it open. What they found instead was a couple of quarts of water stored in the inner flesh of the plant. It is enough water to save someone's life, but it has a bitter flavor. Your thirst has to be real before you'll put up with the taste!

Where does all that water come from in the desert? The cactus has long sharp protective spines. The spines end in a bend toward the ground. They are so designed that they catch the dew and deliver it, drop by drop, to the ground around the plant. There, shallow roots quickly take up the water for storage. The compass barrel cactus also gives directions to the traveler. The cactus produces a light-sensitive chemical that makes it grow more rapidly on the side that gets the least sun. The result is that the cactus grows leaning toward the south.

The compass barrel cactus is more than just a compass for those who are lost in the desert, It also bears witness by pointing to a wise Creator who, we learn from the Bible, provides forgiveness of sins, salvation and spiritual direction through Jesus Christ.

Prayer: Dear Heavenly Father, this world is a spiritual desert. I thank You that You have not left me to wander alone. I ask that You would show me those who are wandering unguided, that I may show them Your guidance. In Jesus' Name. Amen.

An Alga that Flexes Its Muscles

Job 10:8a
"Thine hands have made me and fashioned me together round about;"

A strange alga that has muscle-like tissue illustrates that there are no simple life forms. Protein muscle fibers, say evolutionists, are a later and more advanced development in living things. Yet these scientists also claim that alga are among the first of the living things to have evolved.

The Mougeotia alga contains a structure that performs photosynthesis. This structure is shaped like a flat disc. Researchers found that, depending on the brightness and direction of the light, this structure is turned to control the rate at which photosynthesis takes place. Further study indicates that the structure is actually turned by protein fibers that expand and contract just like our muscles to keep everything under control!

Here we have two very different structures working together. The disc-like structure that carries out photosynthesis could not function very well at all without the "muscle" fiber. If the muscle fiber were to evolve without the disc-like structure to control, the fiber would be useless.

The individual parts of all living things work together like the individual instruments in an orchestra. They're useless without each other. This example shows that individual structures in living things could not have evolved gradually and individually, over long periods of time. They were created to work together in close harmony!

Prayer: Dear Father, only You could have kept track of all the details which were necessary to build the wonderful creation in which we live. Help me to remember Your attention to detail and Your love for me when I am tempted to think that some difficulty is too minor to bring to You. In Jesus' Name. Amen.

Ref: Wagner, C., W. Haupt, and A. Laux. 1972. "Reversible inhibition of chloroplast movement by cytocholasin B in the green alga mougeotia." *Science*, v. 176. p. 808.

The Sawfly

Isaiah 45:18
"For thus saith the LORD that created the heavens; God himself that formed the earth and made it; he hath established it, he created it not in vain, he formed it to be inhabited: I am the LORD; and there is none else."

Most birds and animals leave the beautiful Monarch butterfly alone because it tastes awful. The Monarch butterfly manages to taste so bad by storing some of the juices from the bitter plants it eats. It appears as though the sawfly uses a very similar tactic.

The larvae of the sawfly live in live pine trees. As larvae, they collect a bitter turpentine-like oil from the pine tree. When they are attacked by a predator, they discharge some of the turpentine-like material to form a bitter oily covering around themselves. One wonders if, long ages ago, some sawflies set up a laboratory to research a better defense. Would they have tested various substances on predators to find out just which substances tasted the worst? How would they have figured out how to change their bodies in order to store and discharge the foul-tasting material?

That may sound very silly, but no sillier than supposing that nobody at all designed this system! It is only reasonable to believe that such a dandy design had a designer – one who could control life and apply His great knowledge.

These complex designs are all around us to remind us that there is a Creator. If there is a Creator who wants us to know about Him, He must have told us more about Himself. And He has indeed done this in the Bible, where we learn of Christ's work to bring us to Himself.

Prayer: Dear Lord, I thank You that as the Word, You were made flesh to reveal the gracious mind and heart of God toward us in Your work of atonement. Help me never to take Your work of salvation for me for granted. Amen.

Ref: Eisner, T., J.S. Johnessee, J. Carrel, L.B. Henry. and J. Mernivald. 1974. "Defensive use by an insect of a plant resin." *Science*, v. 184. p. 996.

The Lizard with Hair

I Chronicles 16:8-10
"Give thanks unto the LORD, call upon his name, make known his deeds among the people. Sing unto him, sing psalms unto him, talk ye of all his wondrous works. Glory ye in his holy name: let the heart of them rejoice that seek the LORD."

It sounds like a Hollywood invention: A lizard with hair and suction cups! But the gecko is very real and loaded with dandy designs.

Did you ever wonder how a gecko could walk across the ceiling upside down without falling off? No matter how closely you examine a gecko's feet, you can see no reason for him to stay on the ceiling. It is almost as though he had invisible suction cups on his feet.

It turns out that one must magnify the gecko's little foot pads some 35,000 times under an electron microscope before the secret is revealed. The foot pads are covered with uncountable little hairs that are rather like the fibers on your toothbrush. Each of these tiny fibers has a tiny suction cup on the end. It is estimated that the gecko has at least half a billion suction cups on his feet. No wonder he doesn't fall off the ceiling! If you study the way he walks, you will see that his feet and toes are constructed in a special way so the suction cups easily peel off the surface he is walking on.

If evolution had as much power and intelligence as it would take to design and make all these special features and get them to work together, evolution could easily be witnessed in the laboratory. The only answer that seems to make any sense is that there is an all-powerful, all-wise Creator who has no limits to His creativity!

Prayer: Dear Father in heaven, I praise You for all of Your created wonders. Of all You have created, human beings are the most impressive of the visible creatures. I am fearfully and wonderfully made and I thank You. In Jesus' Name. Amen.

Time to Clean Fish

I Corinthians 15:26
"The last enemy that shall be destroyed is death."

The yellow French angel fish is not the only creature in the fish-cleaning business. Far more important to the general fish population in an area is the Pederson shrimp.

The Pederson shrimp is found in the warm waters of the Bahamas. This beautiful shrimp is partially transparent and has white strips and purple spots. A Pederson shrimp will look for a spot where fish seem to congregate and then wave his antenna in order to attract their attention. Any fish who wants to be cleaned swims over to him and allows the shrimp to climb onto his body. The shrimp goes over the entire fish, cleaning any lesions and removing all parasites, including any in the fish's gills. If more than one fish wants to be cleaned, the others wait patiently in line for their turn.

Researchers studying the Pederson shrimp removed all of the known cleaning shrimp from one particular area to see what would happen. Within a few days the fish population in the area was down dramatically. The few fish who were left had ulcers on their skin and ragged fins.

Even though humanity in its rebellion against God has brought death and decay into the world, God has mercifully provided some relief from these effects for all His creatures. Even humans, the cause of the problem, have been given relief in the saving work of Jesus Christ, who restores harmony between us and the Creator through the forgiveness of sins that He purchased for us on the cross of Calvary.

> *Prayer: Dear Father in heaven, I thank You that You have not abandoned Your creation to the consequences of man's sin. I am most amazed and thankful that You have even provided man, who ruined Your creation in the first place, to have Your forgiveness through Your Son, Jesus Christ. Amen.*

Ref Clark, Harold. 1964. *Wonders of Creation*. Mountain View, CA: Pacific Press.

The Most Improbable Honeybee

Psalm 113:5-6
"Who is like unto the LORD our God, who dwelleth on high, Who humbleth himself to behold the things that are in heaven, and in the earth!"

One reason many scientists reject evolution is that it's impossible to explain even the simplest of the incredible designs we find in creation as the work of a mindless, impersonal force. Imagine the challenge of explaining the many specialized features of the honeybee as the work of no one!

The honeybee has compound eyes that enable it to navigate by the sun even on cloudy days because it has a built-in polarized light filter. The honeybee's antennae contain both their sense of smell and sense of touch. For this reason they must be carefully cleaned. So the honeybees have special grooves in their front legs that are perfectly designed to clean them.

The honeybees also have hairs on their body to collect pollen and baskets on their rear legs to carry it. And they have special glands for producing, shaping and cleaning off wax. When one returns home with news of a good pollen source, the honeybee has a language that it can use to tell the others about it!

Not only is the honeybee a huge collection of special features united together, but they also live in a hive that unites thousands of individuals so that they all work together like a single organism! It simply goes against everything we know from science to suppose that millions of years of unplanned accidents could design and build the honeybee!

Prayer: I thank You, Lord, that some who are called by Your Name work in the sciences. I ask You to protect them, make them bold in their witness to the truth. I know that for You, Who so wondrously made the honeybee, this is an easy task if Your people are willing to serve You above all things. Amen.

Ants Who Garden

Genesis 2:15
"And the LORD God took the man, and put him into the garden of Eden to dress it and to keep it."

Farming and gardening are said by evolutionists to be very advanced activities. They say that primitive humans did not do these things, and that gardening developed only recently in human history. But if there is a Creator, we would expect that He would have taught many kinds of creatures to care for plants or even trees.

Guess what? There are even ants who garden. There is a particular type of fierce ant that cares for the South American bull's horn acacia tree. While the ants don't need the tree for their survival, they do eat portions of it. But they never eat enough to cause damage to the tree. In fact, the ants protect their tree – they snip off vines or other growth that comes too close to the tree, maintaining plenty of growing room for their tree. The ants are aggressive enough to keep other insects or even birds or larger animals away from their tree.

In studying this amazing relationship, researchers have removed the ants from some of these trees. Within two to fifteen months the tree is dead. Without the ants' care, animals eat off all the leaves and surrounding plants overrun it.

Who taught these ants to be gardeners? How could two so very different kinds of life develop such a close relationship? This results in a great mystery for evolution. Without ants you couldn't evolve the tree, and without the tree, the ants couldn't learn to care for it. You can't get around it. Both were created fully formed, and the Creator taught the ants how to care for the tree.

> *Prayer: Father, I see for myself that Your Word is true as I look around the world that You have made. I ask that You would show me what You want me to do as Your creation in bearing witness to the truth of the forgiveness of sins, which is ours through Your Son, Jesus Christ. Amen.*

The Cowboy Lasso Mold

Ezekiel 18:23
"Have I any pleasure at all that the wicked should die? saith the Lord God: and not that he should return from his ways, and live?"

Did you know that there is a mold that captures and eats animals? Those scientists who reject evolution in favor of creation often stress that there are no simple forms of life. Each kind of life is both specialized and very complex. While evolutionary scientists try to arrange living things from simple to complex, creation scientists ask them to let us know when they finally find a "simple" form of life.

At first glance, mold would seem to be a simple form of life. It doesn't sing, dance or write plays. But neither do a lot of people. However, mold – like every other life form – is still perfectly suited for its needs.

Several forms of mold actually capture and eat animals – in this case, a small worm known as a nematode. Some molds grow sticky knobs that trap the worms. But one, known as the "cowboy lasso" mold, grows tiny loops or lassos. Should an unwary worm try to crawl through the loop, the loop swells shut, strangling the nematode. The worm is then digested at the mold's leisure. This is much too complex and specialized for a "simple mold" to have engineered. It is too filled with purpose to be the result of purposeless natural laws.

God does not operate without purpose – not when He created mold, nor when He deals with us. His most earnest desire is for a relationship with you and me through Jesus Christ.

> ***Prayer: I thank You, Father, that while there is no simple form of life, the way back to You forever through the forgiveness of sins in Your Son, Jesus Christ, is simple and already accomplished for us. Help me to better appreciate both Your complexity and Your simplicity. In Jesus' Name. Amen.***

The Cyanide Defense

John 12:32
"And I, if I be lifted up from the earth, will draw all men unto me."

Wouldn't it be amazing if a creature developed a chemical defense that is deadly to practically every other creature but itself – especially if that defense meant exposing itself and its enemy to deadly cyanide?

Yet there is a species of millipede called *Apheloria* that does just this. Apheloria has, on each segment of its body, special glands that produce the chemical needed for defense. When the millipede is attacked by an enemy, it mixes this chemical with a catalyst. The result is a chemical that is a mild irritant plus hydrogen cyanide gas – the same lethal chemical used in the gas chamber. In this defense, the millipede, as well as his enemies, are engulfed in a cloud of deadly cyanide gas. His attacker dies, but the millipede simply walks away unharmed.

This truly amazing defense clearly has a purpose. Even more amazing, if evolution is right, the millipede developed this remarkable defense quite accidentally, and at the same time he became immune to deadly cyanide gas! This defense is completely unlikely to develop without a planner. The millipede's immunity to cyanide makes sense if there was a planner, and it makes no sense at all if there wasn't.

God's fingerprints are all over the creation as He seeks to draw us all to Himself.

Prayer: Dear Father, I thank You that You are a loving God, drawing all to Yourself. I confess that there are times when You have sought to draw me closer, and I would not. For Jesus' sake forgive me, and make Your drawing my greatest desire. In Jesus' Name. Amen.

A Most Unlikely Friendship

Isaiah 55:8
"For my thoughts are not your thoughts, neither are your ways my ways, saith the LORD."

When we discover the unexpected in the creation, we are reminded that reality is much bigger than our small thoughts. That's a hint! God Himself says in the Bible, "my thoughts are not your thoughts, neither are your ways my ways" (Isaiah 55:8).

This means that when we study our world, we must keep in mind that nothing forced God to make things a certain way. He was completely free. While God made an orderly creation, His logic and His creativity are far above ours. Those who believe in a Creator learn to expect delightful surprises in God's world. For those who want to say that the world is a result of mindless natural forces, many of these surprises leave them with little idea of how to explain them according to their system.

One of these surprises is found in the strange relationship between the pagurid crab and the sea anemone. The pagurid crab lives in a shell that is really too small to hold it. So the anemone molds itself to cover the crab and attaches himself to the crab. This gives the crab plenty of freedom of movement because part of his living "shell" is pliable. However, the anemone takes advantage of his position by stealing much of the crab's food. But if the crab is removed from this species of anemone, the anemone will die.

This strange relationship is vital to both these creatures – so far apart on the evolutionary scale. This relationship makes no sense at all unless there is a Creator!

Prayer: Dear Father, let me be comforted by Your love to me when I do not understand why You allow certain things to happen in my life. I know that Your ways and Your wisdom are far above mine. I trust You for Jesus' sake. Amen.

The Hunting Wasp

Job 21:22
"Shall any teach God knowledge? seeing he judgeth those that are high."

What if your refrigerator stopped working . . . in August? It would be a real mess. Without refrigeration, we would have to get fresh food almost every day! The hunting wasp doesn't build refrigerators, but it has solved this problem.

It is during the hottest months of the year that the hunting wasp is most active. However, the hot summer months are not the best time to store food that could spoil, especially in the nursery! Hunting wasps eat a very specific species of caterpillar – and caterpillars spoil easily.

When it's ready to lay its eggs, the hunting wasp will capture a caterpillar and, instead of killing it with one sting, will carefully paralyze each of the caterpillar's 13 segments one at a time. It then carefully bites the base of the caterpillar's brain, not to kill it but just to keep it permanently disabled. Then it lays one egg on the caterpillar. As its young grow, the wasp continues to provide specially prepared caterpillars – just enough to keep the young fed and not so many that there will be any spoiled food in the nest.

The hunting wasp's system seems to be based on far more knowledge than a wasp would be expected to discover all by itself – especially since the cost of failure is the death of the next generation. This certainly looks more like the work of an intelligent, personal Creator than the work of mindless chance!

Prayer: Dear Lord, while I stand amazed at the abilities You have given Your creatures, help me to realize at the same time that all my abilities are from You. I have not made them, nor do I own them – You do. Move me, by Your love for me, to use my abilities for Your purposes. Amen.

Ants Who Live with Company

Job 37:14
"Hearken unto this, O Job: stand still, and consider the wondrous works of God."

Cooperation between two different species who depend upon each other for survival is known as symbiosis. This is remarkable enough, but what about cooperation between two different species that don't depend upon each other for survival? This certainly defies that foundational pillar for the theory of evolution – the survival of the fittest.

Examples of these types of cooperation, found abundantly in nature, have caused a growing number of scientists to listen to arguments for intelligent design – and even abandon mindless evolution altogether. Our example today concerns two very different ant species that live in British Guiana. One species, a large brown ant, lives together with a small black ant. They actually breed and live separately, but in times of danger, they work together as a well-trained army.

At first threat to the anthill, the smaller black ants begin the defense. However, if the black ants cannot hold the intruder, the larger, stronger brown ants join in the defense of the hill. Neither ant needs the other, but together each offers the other some benefit. This is called altruism.

One cannot picture ants sitting around a table to work out this arrangement – they must have been created for this kind of relationship by a good and wise Creator. His plan to help these ants, amazing as it is, is nothing compared to His plan of salvation for us through the forgiveness of sins that is in Christ Jesus!

Prayer: Dear heavenly Father, I thank You that Your greatest work is Your plan of salvation, restoring us undeserving sinners to Yourself through the forgiveness of sins earned by Christ on the cross. I ask that Your plan of salvation would always be central in all my thoughts, despite the distractions of the world. In Jesus' Name. Amen.

Who Is Against Evolution?

Job 21: 7, 14
"Wherefore do the wicked live, become old, yea, are mighty in power?...Therefore they say unto God, Depart from us; for we desire not the knowledge of thy ways."

Lots of folks would like you to believe that only ignorant, backward people reject evolution. But is that really the case? The fact is that the case for evolution is so weak that many scientists who cannot even be called friends of Christianity reject it on scientific grounds!

Back in 1981, Dr. Colin Patterson of the prestigious British Museum of Natural History shocked the scientific world. He told scientists at the famed American Museum of Natural History in New York that he'd been examining anti-evolutionary ideas for about 18 months. It finally struck him that, although he had been working on evolution for 20 years, he could not find one fact about evolution that he really knew.

Dr. Patterson said he had asked scientific colleagues at other institutions whether they knew anything about evolution to be actually true. After a lot of silence at several meetings, one fellow scientist finally spoke up at one meeting and said, "I do know one thing – it ought not to be taught in high school."

Dr. Patterson concluded his talk at the American Museum of Natural History by saying that he finally "woke up and realized that all my life I had been duped into taking evolution as revealed truth in some way." You see, those who reject evolution are in a lot of good, educated company!

Prayer: Lord, I mourn for the many who try to avoid You by hiding in stories about evolution, especially since I know that they cannot avoid coming face to face with You at the judgment. Despite the fact that many of them have set themselves as enemies of Your people, I pray for them and ask that they may not avoid coming in repentance to You before it is too late. Amen.

Is Evolution Simply Change?

I Corinthians 1:20
"Where is the wise? where is the scribe? where is the disputer of this world? hath not God made foolish the wisdom of this world?"

Many people think that evolution is simply any change. Developing a new breed of dog is often referred to by many people as proof for evolution. But is it really? Those who claim that the domestic breeding of crop or animal species has anything to do with evolution take their cue from Darwin, who – although he had no degree in science – made the same claims.

Evolution is not simply change like what we see in domestic breeding programs. You might say that it's easy for a creationist to say this, but let's allow a scientist who is neither a creationist nor a Christian make the case for us. In his book *Evolution: A Theory in Crisis*, Dr. Michael Denton explains that evolution is in crisis for the simple reason that biology offers no evidence that evolution happens. Since science relies on evidence, he is really saying that evolution is not science at all.

Denton, who has spent much of his scientific career studying evolution, adds insult to injury by stating that "Despite more than a century of intensive effort on the part of evolutionary biologists, the major objections raised by Darwin's critics … have not been met."

In other words, evolutionary scientists have not been able to answer creationist challenges. Or, as we have been saying all along, the Bible does offer an intelligent alternative to evolution!

Prayer: Dear Father, I thank You for those honest scientists who, even though they do not believe in Christ as their Lord and Savior, are still honest enough to identify a falsehood. Today I especially ask that You work in their hearts so that they may seek and find their Savior, Jesus Christ. In His Name. Amen.

What Is Faith?

Hebrews 11:1
"Now faith is the substance of things hoped for, the evidence of things not seen."

What is faith? Is faith something we believe despite the facts, as many would tell us? Is faith a blind leap into the dark?

Hebrews 11:1 says, "Now faith is the substance of things hoped for, the evidence of things not seen." Our Christian faith is not a blind leap into the dark. God has given us minds, and He honors that gift by giving us reason to believe that Jesus Christ is the promised Savior. Consider: God gave us hundreds of prophecies about this Savior, saw to it that each one was fulfilled in the life of Christ, and then made sure that people actually saw that the prophecies were fulfilled. That is indeed powerful evidence – more than reasonable cause for us to believe His Word.

Our argument from Old Testament prophecy also shows that God does not expect faith to be a belief in something despite the facts. The world that the Bible talks about – the one God made as recorded in Scripture and the one science studies – is the real, material world in which we live. Because we know that the Word of God is true, we have every right to expect the facts about this world as related in Scripture to agree with facts established by experience.

Faith, as the substance of things hoped for and the evidence of things not seen, is better grounded in God's Word and the world we know than is the so-called science of evolution – which is truly a faith in things unseen.

> **Prayer: Dear Father in heaven, I thank You that Your Word, the Bible, is true and trustworthy. Help me to use the Bible in a way that reflects its true importance, and teach me as I study. In Jesus' Name. Amen.**

The Faith of the Evolutionist

Hebrews 11:3
"Through faith we understand that the worlds were framed by the word of God, so that things which are seen were not made of things which do appear."

Many people think that faith is strictly a religious matter and concerns unprovable issues such as angels, heaven and, of course, belief in the creation story. They further feel that belief in evolution is more reasonable, since it is based upon hard and proven facts. But is this true?

Faith is what we add to the evidences we have in order to provide our worldview. The less evidence there is, the more faith we need. Evolutionary scientists often admit that they, too, interpret the world in the context of their faith. Their faith is that everything and everyone got here by means of evolution.

Prof. L. H. Matthews, a well-recognized evolutionist, was honored by being asked to write a new introduction for the 1971 edition of Darwin's *Origin of Species*. In his introduction – speaking of evolution – he admitted, "Most biologists accept it as though it were a proven fact, … although this conviction rests upon circumstantial evidence, it forms a satisfactory faith on which to base our interpretation of nature."

You see, this is nothing more than what Bible-believing Christians do when they understand the world in the context of their faith. As Christians, we should not be intimidated into thinking that the faith of the evolutionist is somehow superior to ours for understanding the world!

Prayer: Lord, the devil is not called the "Accuser" for nothing. He even tries to make the faith You have give me into a sin! Rather than being intimidated because I believe Your Word, I ask that You would give me a bold faith which does not shrink from speaking Your truth in love. Amen.

Micro-Marvels of the Human Eye

Matthew 11:4-5
"Jesus answered and said unto them, Go and shew John again those things which ye do hear and see: The blind receive their sight, and the lame walk, the lepers are cleansed, and the deaf hear, the dead are raised up, and the poor have the gospel preached to them."

The tiny computer chip – the basis of modern computers – is a thin wafer of silicon only about 7 millimeters across – about one-fourth of an inch. This tiny chip might have the equivalent of 100,000 transistors built into it. Despite its size, hundreds of connections might be attached to it. The design of the computer chip is a marvel, surpassed only by the amazing technology that can build it.

If you were to find a silicon chip laying in the silicon sand at the beach (which would be quite an accomplishment, because they are so small!), you would never be able to convince even one engineer that the chip had formed through the chance movements of all that silicon-based sand!

Now consider the human eye. The retina at the back of the eye is a very thin membrane, much thinner than even the clingiest food wrap. This retina contains no transistors but much more sophisticated photo-receptors, each of which is a high-gain amplifier. The retina doesn't have 100,000 of these, but rather it has 200,000 for each square millimeter of retina!

But the greatest wonder of all is how any scientist can say that your miraculous eye is the result of unguided genetic mistakes that, just by chance, resulted in the eye. Indeed, God's fingerprints are all over the creation, just as our human work is seen in the computer chip. How can one who can design only the computer chip dare question the One who can design the eye – and so much more!

Prayer: Dear Father in heaven, just as man's most creative technology is nothing compared with what You have made, so all of our best efforts are like filthy rags before You. For this reason, I must rely entirely on what Your Son, Jesus Christ, has done for me so that I may have the forgiveness of sins. Thank You. In His Name. Amen.

The Incredibly Sensitive Eye

Job 19:25-27
"For I know that my redeemer liveth, and that he shall stand at the latter day upon the earth: And though after my skin worms destroy this body, yet in my flesh shall I see God: Whom I shall see for myself, and mine eyes shall behold, and not another; though my reins be consumed within me."

The human eye is so incredibly sensitive that it can actually detect a single particle of light – a photon.

The truth is, the human eye is so sensitive that were it not for special features inside the eye that process the billions of pieces of information coming into the eye every split second, we would be overwhelmed. While the eye can detect even one photon of light, it will not pass an image on to your brain until at least six photons strike in the same area of the eye. If this weren't the case, on a dark night we would see nothing but static, since less than six photons could not be focused into an image and would appear to us as just static.

This special provision makes one wonder if the eye was not designed by an all-wise Creator. After all, how could mindless chance and mutations know about the basic laws of physics that control light's behavior? The range of the eye's sensitivity is also a million times greater than our modern photographic films, providing us with a dynamic range of 10 billion to one. While the greatest sensitivity is needed on dark nights, an internal control in the eye reduces that sensitivity for bright daylight.

Science doesn't rule out God at all! Our knowledge of the eye's working demands the conclusion that we are the creation of a wise and powerful God!

Prayer: Dear Father, while I am thankful for and amazed at all of the senses and abilities which You have given me, I am even more filled with wonder that this body shall be raised in Resurrection, free from all imperfection, to live forever with You. I thank You for Your grace in Christ Jesus. Amen.

Have Evolutionists Found a Bad Design in the Eye?

Genesis 1:31
"And God saw every thing that he had made, and, behold, it was very good. And the evening and the morning were the sixth day."

Evolutionists, trying to answer creationist arguments, have suggested that there is an "error" in the design of the eye that any wise Creator would not have made. That "error" in design, as evolutionists call it, is that the retina of the mammal's eye is "inside out." The light entering the eye passes through other eye tissue before hitting the photoreceptors. But is this really an error in design?

Dr. Joseph Calkins, a professor at Johns Hopkins University and a creationist, points out that the photoreceptors in mammals' eyes need the extra tissue. It provides nutrients to the retina. This is crucial, because the eye's receptors have a very fast rate of metabolism – they live out their entire lives in only about seven days!

If you have ever looked at the sun and then experienced an after-image, you have probably burned out some of your photoreceptors. However, because your photoreceptors have such a fast rate of metabolism and those extra nutrients from the eye tissue, the damaged receptors are all replaced within a few days! Besides, as Dr. Calkins points out, the tissues that lie between the light source and your retina are packed in so tightly that they are separated by less than the wave-length of light, making them completely transparent!

It seems that the evolutionists' claim that the eye is poorly designed – and thus a product of chance rather than a Creator – was based on their ignorance of how the eye needs to work. Again, the evolutionary argument falls in the light of scientific knowledge, and once more we see the witness to our Creator!

> ***Prayer:*** *Dear Lord, in order to avoid his own guilt, man would rather fault You – even claiming at the same time that You don't exist! I thank You that You have given me faith in Your love and forgiveness for me. I ask that my trust will always be in You and never in myself. Amen.*

The Eye's Computer

Matthew 6:22-23a
"The light of the body is the eye: if therefore thine eye be single, thy whole body shall be full of light. But if thine eye be evil, thy whole body shall be full of darkness."

Computer scientists, in trying to build a machine that sees like we do, have been studying how the eye works the miracle of vision. What they are discovering provides more scientific evidence that evolution is nothing more than a modern fairy tale for adults.

Computer scientists have learned that before one single image is ever sent to your brain, each cell of your retina must perform a huge number of calculations. Each second, every cell of your retina performs 10 billion calculations! And you thought you couldn't do math!

What's more, these are not simple calculations. Dr. Joseph Calkins, professor of ophthalmology at Johns Hopkins University, estimates that the fastest computer in the world would require many hundreds of hours to do what the cells in your retina do each second! It is easy to understand why a scientist who studies eyesight cannot accept the idea that the eye simply evolved.

The eye is witness to more than the fact of the Creator. The clear vision of His involvement with humanity begins when Jesus Christ gives us clear spiritual sight in cleansing us from our sins. Just as many people must wear glasses to see clearly, because of our sin, we cannot clearly see our Creator who seeks us. Yet He is not far away, and He desires more than anything else that we come to Him in the Name of Jesus Christ. In Christ you will see miracles greater than vision!

Prayer: Dear Heavenly Father, grant me clear sight to see all things, spiritual and earthly, in the true light that You provide for me in Your Word. Let me see Your working in my life through Your Word so that my whole person radiates Your light. In Jesus' Name. Amen.

Could Science Make an Eye?

Psalm 139:14
"I will praise thee; for I am fearfully and wonderfully made: marvelous are thy works; and that my soul knoweth right well."

Scientists have tried to tackle the job of creating a machine that is able to see as well as the human eye. In the process, they are gaining a new appreciation for the wonderful gift of sight.

They have learned, first of all, that no computer chip can be made today that could begin to do what the retina does. But, if it could be made, it would have to be something like half a million times bigger than your retina. One computer scientist has estimated that a computer chip that could begin to do what your retina does would have to weigh in at about 100 pounds!

The retina, which is something like a small slip of clingy food wrap, weighs less than a gram. It occupies only 0.0003 inch of space. On the other hand, the scientists' theoretical "seeing" chip would fill 10,000 cubic inches. And while your retina operates on only 0.0001 watt of power, the synthetic "seeing" chip would require 300 watts of power and a cooling system. Even with all this, it couldn't see very well. It would be able to resolve a square area of only about 2,000 units of vision – called pixels – while your eye can resolve five times that much! Such a chip would have the equivalent of about 1 million transistors, while your retina has the equivalent of 25 billion transistors!

Truly, modern science offers elegant testimony to the fact that the eye could not have been produced by evolution – and that it could have been created only by a very wise Creator!

Prayer: Truly, dear Lord, I am fearfully and wonderfully made! I thank You for all the senses You have given me. Even though they may not all be perfect, I ask that You would perfect them by leading me to use them to Your glory. In Jesus' Name. Amen.

One Designer

Romans 1:20
"For the invisible things of him from the creation of the world are clearly seen, being understood by the things that are made, even his eternal power and Godhead; so that they are without excuse:"

If you give a design problem to ten different engineers to solve, chances are you will not get ten different solutions. In fact, you are likely to end up with ten solutions that look very much alike.

Now suppose 1,000 years from now someone found your ten sheets of problems, each with the design solutions offered by the ten engineers. Since they are practically the same, this person might conclude that each of the solutions was offered by the same person. He would account for their differences by saying that as each solution was tried, better ideas were developed. He would conclude that the solution most familiar to his own day was the final and most advanced solution.

When evolutionists compare creatures like humans with apes, noting the similarities, they are falling into the same trap as our theoretical researcher from the future. Certainly there are some similarities between humans and apes. However, the structure of the human eye is most similar to the eye of the octopus! The general pattern of the functional structure of hemoglobin in human blood differs by only one atom from the chlorophyll found in plants. And the protein in human milk is most like the juice of the soybean plant.

None of these comparisons offers any evidence for evolution. What they illustrate is that any good designer will solve similar problems with similar designs – they offer evidence that the creation is the work of a wise and powerful Designer!

Prayer: I thank You, dear Father, for Your continued love for man despite our stubborn perversity. I thank You that You sought me when I did not seek You and that You first loved me so that now I might love You. In Jesus' Name. Amen.

Amazing Water

Job 37:10
"By the breath of God frost is given: and the breadth of the waters is straightened."

According to evolutionary theory, all matter and energy were the result of a huge explosion called the "Big Bang." The laws that matter and energy must follow are also the result of that same great explosion. As a further result of these beliefs, evolutionists are convinced that things behave the way they do quite accidentally.

The first question that comes to mind is: How could an accident have produced the seemingly careful designs we see in the way certain important materials behave? Consider water, for example. Water is essential to life. Water, which is the basis of our blood, carries dissolved food to the deepest cells in our bodies along with oxygen so that our cells can live. Water dissolves the wastes and behaves in just the right way so that other organs can remove those wastes from our bodies. Is it an accident that *only* water, the very same material basic to the materials of life, also can do all these other unique jobs?

Water also refuses to act like most other materials. For example, we all know that when any material is turned from a liquid to a solid, it becomes more dense and therefore heavier. But when water freezes into ice, it doesn't do this; it gets lighter. If it got heavier, ice would sink in our northern lakes when it formed, and they would quickly freeze solid, killing all life in them.

The unique properties of water are only a few of the millions of so-called accidents that had to happen "just so" – in harmony with millions of other details – in order to make life possible. All by itself, simple water testifies to a wise Creator.

Prayer: Dear Father, I thank You for the wonderful way in which You have designed blood to work. But most especially do I thank You that Your Son, Jesus Christ, shed His blood on the cross for the forgiveness of my sins. Let me never take that wonderful blessing for granted in any word, deed or thought. In His Name. Amen.

The Marvel of Life

Job 12:9-10
"Who knoweth not in all these that the hand of the LORD hath wrought this? In whose hand is the soul of every living thing, and the breath of all mankind."

We've all heard it said that we are made up of about 98 cents worth of chemicals. However, the living matter that makes up our bodies is very different from non-living matter.

Our growing understanding of biochemistry continues to close doors to evolutionary theory and open doors to the Bible's claims about the origin of life. Many evolutionary scientists are admitting that to say life is simply the right kind of chemistry is like saying that a computer is simply the right kind of plastic, metal and electricity.

Growing knowledge of the intricate information codes in living things has led many evolutionists to search for other theories, and some have even considered creation. Creationism is not growing in influence because it has good funding – it certainly doesn't. Creationism is gaining followers because it is becoming increasingly clear that no known principles can account for the development of living things. If science is proving anything, it is proving that life is impossible and should not exist!

This is exactly what the Bible tells us. Without God's direct activity within the world, life could not have come about – only God can be the source of life. This clear fact should steer all of us to the rest of God's message in Scripture. There, God tells us that life is more than just food and drink – the material things of life. Full and complete life that satisfies the whole being is found only in Jesus Christ.

Prayer: Dear Lord, only in You can I or anyone else find true life – life which never ends. Do not allow this truth to be clouded in my life or witness by the cares and concerns of the material needs of life here on Earth. Amen.

How Old Does the Bible Say the Earth Is?

Genesis 5:3-5
"And Adam lived an hundred and thirty years, and begat a son in his own likeness, after his image, and called his name him Seth: And the days of Adam after he had begotten Seth were eight hundred years: and he begot sons and daughters: And all the days that Adam lived were nine hundred and thirty years: and he died."

We have all heard or read about a scientist from some university declaring that a certain rock or fossil is millions of years old. But what do these pronouncements do to the Bible's history that begins with creation? Has modern science proven that the Bible's history is inaccurate?

Nowhere in the Bible do we find an exact statement that says precisely how many years old the Earth is. However, the Bible is filled with statements that, if added together, can give us a fairly accurate age for the Earth. You see, in ancient times people did not receive pretty picture calendars from the local bank. Their calendars were much different from ours today – but they were very accurate. These calendars are recorded in the Bible.

Many people find little use for the lists in Scripture that say so-and-so begat so-and-so, who begat so-and-so when he was 70 years old. But these are not lists. They are the ancient calendars, preserved for us today in the Bible by God. Even the Chinese preserve their calendar by genealogies. If we chain all the "begats" together and add up the results, we find that the Earth cannot be much more than 6,000 years old. Even if there might be a few gaps between some of the "begats" (which seems unlikely), the Earth still cannot be much more than 7,000 years old.

There are actually more scientific dating methods that support this young age for the Earth than support the most ancient ages. The Bible's history stands unchallenged by modern science!

Prayer: Dear Father, I thank You that Your Word is trustworthy. Forgive me when I begin to think that our modern age has more truth than previous ages, and especially more than Your Word. In Jesus' Name. Amen.

Scientific Dating

Psalm 10:4-5
"The wicked, through the pride of his countenance, will not seek after God: God is not in all his thoughts. His ways are always grievous; thy judgments are far above out of his sight: as for all his enemies, he puffeth at them."

The subject of scientific dating has to do with scientific methods used to try to figure out how old the Earth is. However, you should know that what evolutionary scientists aren't telling you about their methods changes the whole picture!

You see, there are over 100 different methods that may be used to date the Earth or the universe. For example, if we measure the amount of dissolved aluminum in the oceans and then measure the rate at which aluminum is being carried into the oceans by the world's rivers, we can scientifically "prove" that the Earth is not more than 100 years old! Now that "scientific" finding is silly. The truth is, there are half a dozen scientific methods for dating the age of the Earth that "prove" that it is 1,000 years old or less. Fewer than a dozen methods show that the Earth is over 1 billion years old!

When mathematicians encounter such a wide range of results, especially when some of the results at the extremes of the range are clearly wrong, they begin to eliminate equal numbers of results at both ends of the extreme. If we do that with the results of modern scientific methods, no method would be left that shows the Earth to be much more than 100 million years.

The most amazing things is that fully 15% of all the dating methods cluster around an age for the Earth of between 6,000 and 10,000 years. So the scientific evidence, fairly considered, fits in very nicely with the age for the Earth of around 6,000 years that is found in the Bible.

Prayer: Dear Father in heaven, You have seen man, throughout his history, deciding to follow those explanations which take him further from You. This makes me all the more thankful that Your grace through Jesus Christ has found me. In His Name. Amen.

How Deep Is the Moon's Dust?

Genesis 1:16
"And God made two great lights; the greater light to rule the day, and the lesser light to rule the night: he made the stars also."

Did you know that the actual dust on the surface of the moon is thousands of times less than expected by those who think the Earth is billions of years old?

All of us know that if something has a lot of dust on it, it probably hasn't been cleaned in a long time. If something has very little dust on it, it may have been just cleaned – or it might be brand new. Since there is no such thing as a moon cleaner, if our moon doesn't have much dust on it, it must be fairly new.

Before American astronauts landed on the moon in 1969, space scientists were worried that a moon landing would be impossible. By that time, scientists knew how much dust there was in space, and they knew how fast this dust would accumulate on the moon. Since they figured that the moon was more than 3 billion years old, they reckoned that there could be as much as 150 feet of soft dust on the moon – so deep and so soft that a manned lander might sink into the dust and never be heard from again. For this reason, they designed the lunar lander with large pads to support the machine on the soft dust.

But we all know what happened – there wasn't even enough dust to plant the American flag. The flagpole had to be supported with rocks! This is exactly what creation scientists told them they would find, since the moon has been accumulating dust for only a few thousand years and not billions or even trillions of years!

***Prayer:** Father, I thank You for the beauty of the sky, especially the moon, which dominates the night sky. Here I see Your power in the many objects You have created. Help them to remind me that this Earth is not the only world where I shall live, and as a result seek Your Word in Scripture that I might be better prepared for the new heavens and Earth. In Jesus' Name. Amen.*

Cave Mysteries

II Samuel 22:31
"As for God, his way is perfect; the word of the LORD is tried: he is a buckler to all them that trust in him."

Have you ever been in a cave? If you have, perhaps someone has told you that the rock formations hanging from the ceiling – called stalactites – are thousands or even tens of thousands of years old. Scientists tell us that stalactites take 100 years, on the average, to grow one inch. But just how accurate is this figure?

Stalactites grow where water seeps through limestone rock, dissolving limestone in the process. When this water containing dissolved limestone emerges from the roof of a cave, it hangs for a moment. In a current of air, some of it evaporates, causing the limestone to deposit. Finally, the remaining water drops to the floor of the cave and continues to evaporate and deposit. The deposit on the floor is called a stalagmite.

A concrete railroad bridge in Wooster, Ohio had a stalactite growing under it that was over 12 inches long! Had the railroad bridge been standing for more than 1,200 years? Obviously not – in fact, the bridge had been cleaned of stalactites only 12 years before! Nor is this situation unusual. More than 300 stalactites were counted growing under bridges in just this one city, and stalactites are not hard to find under concrete bridges in most cities.

We take so many things at face value that evolutionists tell us about the world, even if they contradict the Bible. Before we start taking human words on faith, we should take at face value what the Creator says in the Bible. He knows more than all scientists put together and is to be trusted above any human teacher!

Prayer: Lord, men find so much to be afraid of and I confess that I, too, fear too much. Yet mankind runs away from You, the One Who could calm all their fears. Help me to see that when I am afraid, I am running too. Forgive me, and let my trust in You for all things become stronger. Amen.

Were You Once a Fish?

Romans 1:25
"Who changed the truth of God into a lie, and worshipped and served the creature more than the Creator, who is blessed for ever. Amen."

In 1866 a German scientist and strong supporter of Darwin, Ernst Haeckel, got into trouble for faking evidence. He supposedly showed that developing embryos go through the earlier evolutionary stages of their species. The scientific community investigated his so-called discovery and found that he had faked his drawings. Unfortunately, textbooks today – well over a century later – still present this faked evidence in support of evolution.

As a result, many people today still believe this false scientific theory. One of the most widely used "evidences" used to support this idea today is the claim that the developing human fetus has "gill slits" at one point in its development. In fact a United States ambassador used this argument in support of abortion in a recent issue of National Review.

The truth is, the developing fetus has folds of skin around its neck that contain the developing organs found in the neck. These folds have nothing in common with gills. While no scientist today accepts this theory – and it is even refuted in most encyclopedias – it can –be found in textbooks used by high school students! Why? It would seem that few scientists have the courage to challenge this idea they know is wrong because it might make them appear to oppose evolution! But even worse is that this anti-scientific idea continues to be used as a "scientific" justification for abortion.

Yes, it's true. The wrong view on origins can result in some disastrous "science" as well as human suffering!

Prayer: Dear Father in heaven, it is especially easy even for Your people to fall into an inadvertent worship of the creation in place of the Creator in our materialistic age. Help me to root this sin out of my life. In Jesus' Name. Amen.

Are Birth Defects God's Plan?

Genesis 3:8
"And they heard the voice of the LORD God walking in the garden in the cool of the day: and Adam and his wife hid themselves from the presence of the LORD God amongst the trees of the garden."

Most creationists are pro-life because they see conception and birth as one of the clearest evidences of God's original role as Creator. Scripture is filled with examples showing that God not only created the heavens and Earth in six days, just as reported in several places in Scripture, but that He is still intimately involved with His creation.

In other words, God did not create the heavens and Earth and then leave the Earth to spin off on its own in space on automatic pilot as mindless natural laws took over. This means that there is no such thing as chance. Can birth defects, therefore, be ascribed to God's will?

The Bible also tells us that God created a perfect creation, with no sin, death or defect of any sort. It was we humans who changed the creation when we sinned by abandoning God. He never left us. Unfortunately, our sinful condition brings its own results – including more sin, genetic defects and death. Jesus' disciples once asked Him about a man born blind, "Master who did sin, this man or his parents?" Jesus' answer shows us God's intimate involvement in human life. He told the disciples that the man was born blind to show forth the works of God. Then Jesus miraculously healed his blindness.

There is no worse thing for us than sin and its results now and forever. But the lesson is if we allow God, our Creator, to work, there is nothing so evil that it cannot, in His hands, show forth His mighty works!

Prayer: I lament, Lord, that because of sin You no longer walk and talk with us as You intended – as You did with our first parents before the fall. But I thank You that through Your Son, Jesus Christ, we are restored to a relationship with You where You are with us each day and where You would speak with us each day in Your Word. Amen.

Darwin's Puzzle

John 18:38a
"Pilate saith unto him, What is truth?"

How do you know that what you think you know is really true? Charles Darwin wondered just that, and the answer to his question sheds a lot of light on the origins debate today.

Since Darwin had no formal training in science, he made his case for evolution from philosophy, not from science. Philosophy and theology, after all, was the area in which he was trained. This background led him to ask a very important question. In Darwin's own words, ". . . the horrid doubt always arises whether the convictions of man's mind, which was developed from the mind of lower animals, are of any value or at all trustworthy. Would anyone trust in the convictions of a monkey's mind, if there were any convictions in such a mind?"

In other words, what Darwin was saying was that if his theory was true, it was the product of a mind not much greater than a monkey's. And who, including Darwin himself, could trust such a mind? The only way in which human thoughts might be separated far above the animals is if creation is true. Either way, the logical conclusion of Darwin's puzzle is that creation is true and evolution is untrustworthy!

It is no accident that as the teaching that humans came from lower animals has grown, the number of people who act like animals has also grown. Darwin's own statement seems to show the twisted logic that results from evolution.

Prayer: Dear Father, I pray that our age which asks with Pilate, "What is truth?" may become more open to Your Word, the only truth, as it learns to despair of man's wisdom. In Jesus' Name and for His Glory. Amen.

Robert Boyle: Creation Scientist

Ephesians 2:10
"For we are his workmanship, created in Christ Jesus unto good works, which God hath before ordained that we should walk in them."

We often hear today that no true scientists are creationists because science disproves creation. The truth is, most of the founders of the various disciplines of modern science were creationists, as are a great many scientists today.

One of the most famous scientists of all time was Robert Boyle. Boyle, who lived at the close of the 17th century, did pioneering work in both chemistry and physics. Perhaps you are familiar with "Boyle's Law," which relates the pressure, temperature and volume of gas. "Boyle's Law" is studied by every high school student today.

Boyle was also a completely dedicated Christian. He contributed a great deal of his money to the work of translating the Bible into languages that did not yet have the Scriptures. Boyle was also very concerned about those who, in his own day, thought science could prove the Bible wrong. These people were the forerunners of modern evolutionists. In order to combat this idea in science, Boyle, with others, founded the Royal Society of London as one of the earliest modern creationist organizations. He also funded the famous "Boyle Lectures" in his will, stipulating that these lectures were to offer scientific viewpoints that upheld the truth of the Bible.

Boyle called science a religious task that has the job of learning more about the workmanship of the Creator.

> ***Prayer:*** *I thank You, Lord, for the scientific contributions and the example of faith provided by many of the great scientists like Robert Boyle. I ask that You would stir the faith and talents of Your people today so that we may see many more servants like Robert Boyle in Science. In Jesus' Name. Amen.*

Sir Isaac Newton

Proverbs 2:6
"For the LORD giveth wisdom; out of his mouth cometh knowledge and understanding."

If Isaac Newton or someone else had never made the discoveries Newton made, our world would be very different today. He is one of the most important scientists in all of history. What many people don't know is that this great scientist was a creationist. He actually wrote more Bible commentaries than he did scientific papers.

Newton is best known for his discoveries of the law of universal gravitation and the three laws of motion. Newton also built the first reflecting telescope, and he developed calculus into the branch of mathematics it is today. He also researched the nature of light and explained how white light is made up of many different colors.

Newton was an avid student of the Bible and wrote papers defending Bishop Ussher's dating of the Earth at about 6,000 years. He also defended the literal six-day creation. Although Newton did not personally accept the doctrine of the Trinity, he did take on the atheists of his day, offering strong defenses for the biblical view of history. He was one of the first creation scientists to suggest that most of the sedimentary rocks of the Earth are the result of the Genesis Flood.

Next time you hear someone say that science and the Bible don't mix, think of Sir Isaac Newton.

> ***Prayer: Dear Father in heaven, give me a mind which seeks to be closer to You and to learn more about how You have made all things. Though I may not have the understanding of a scientist, I desire more reason to praise You. In Jesus' Name. Amen.***

James Clerk Maxwell

I Corinthians 2:5
"That your faith should not stand in the wisdom of men, but in the power of God."

Albert Einstein said that James Clerk Maxwell made greater contributions to physics than anyone except Isaac Newton.

Maxwell developed complex theoretical and mathematical explanations for all the forces in the universe except gravity and nuclear forces. He also made scientific contributions in the fields of thermodynamics and mathematics. In other words, Maxwell was a scientist of gigantic proportions who remains greatly respected today.

By today's standards, Maxwell would be called a "fundamentalist." Maxwell lived at the same time as Charles Darwin and was very aware of evolutionary theory. He felt strongly that evolution was anti-scientific and wrote a powerful and important refutation of evolutionary writings. He also offered a very careful mathematical refutation of the theory that the solar system had evolved from a cloud of dust and gas.

The great scientist Maxwell believed that Jesus Christ is the Savior God has provided to deliver humanity from the results of sin – including eternal death. A writing of his, found after his death, states that the motivation for his work was that God had created all things just as Genesis says. And since God has created humans in His image, scientific study is a fit activity for one's lifework.

Prayer: Dear Heavenly Father, I pray today for the work of those in science who are convinced that You are indeed the Creator as described in Genesis. Though they are opposed by men, bless their work and move more of our Christian young people to follow in their footsteps. In Jesus' Name. Amen.

More Butterfly Colors than You Can See

Psalm 27:4
"One thing have I desired of the LORD, that will I seek after; that I may dwell in the house of the LORD all the days of my life, to behold the beauty of the LORD, and to enquire in his temple."

The beautiful wing of the butterfly has a lesson to teach everyone who believes living things were created by chance. The design of the butterfly wing involves much more than just intricate complexity. In addition to showing a knowledge of flight, each wing design is the product of very precise specifications built around the specific wave-lengths of visible light.

The iridescent colors you admire on a butterfly's wing are created by the scales of the wings. Each square centimeter of wing has tens of thousands of these scales, each attached to the wing by tiny stems and overlapping each other like cedar shakes. Each one of these scales was a living cell until a day or two before the butterfly emerged from its pupa. Each tiny scale is made of a vertical and horizontal framing within which are found various sacks of pigment hanging from the framework.

Butterfly wings that seem to glow with iridescent blues and greens have scales with tiny lattices and ribbed walls that are designed to cause interference patterns in light waves within the 300-700 nanometer range – exactly the range humans see as color. That interference pattern is what our eyes interpret as iridescence.

It takes a good knowledge of physics as well as micro-architecture to design and build an iridescent butterfly wing! Science clearly teaches us that such ability and knowledge does not come from chance. Let's be bold to admit the truth – there is a Creator!

Prayer: Dear Father in heaven, a child can see the beauty of Your handiwork in the wing of the butterfly. Through my witness, help those who have been fooled by evolution to see You in Your creation and our salvation in our Savior, Jesus Christ. In His Name. Amen.

Surprise of the Fire Salamander

II Peter 3:3-4
"Knowing this first, that there shall come in the last days scoffers, walking after their own lusts, And saying, Where is the promise of his coming? for since the fathers fell asleep, all things continue as they were from the beginning of the creation."

The fire salamander is a six- to eight-inch-long amphibian found throughout Europe. While it has some amazing abilities, legend gives this potentially dangerous creature even more abilities.

According to medieval legend, the fire salamander could not only live in a fire, it could thrive in the fire. If the creature touched a tree, people thought that the fruit on the tree would be poisonous. Of course, it's highly doubtful that any of these legends will be proven factual by science.

According to a 17th-century naturalist, the fire salamander is also able to squirt a white liquid from its skin to repel attackers. In 1900, a French biologist pooh-poohed this claim, although he admitted that the creature's skin did secrete a powerful nerve toxin that can be fatal if swallowed. Naturalists accepted this description until very recently. Researchers have now shown that the fire salamander does indeed spray its poison at attackers – and can hit them up to seven feet away! They have also established that the salamander seems to aim for the face – a fact known well by one researcher who was accidentally blinded for 20 minutes by one spray!

In their pride, modern humans seem to feel that any knowledge older than a couple of centuries is untrustworthy. Confirmation of older research as well as much of the biblical history should offer a humbling lesson to modern humanity's know-it-all attitude!

Prayer: Dear Lord, I live in a proud age which listens to few before its time. Fill me with humility, that I may rely less on myself and more on Your instruction and power in Your Holy Word. In Jesus' Name. Amen.

They Talk to the Trees

Colossians 1:16b-18
"...all things were created by him and for him: And he is before all things, and by him all things consist. And he is the head of the body, the church: who is the beginning, the first born from the dead; that in all things he might have the preeminence."

In Costa Rica's tropical forests there are ants who live inside young piper trees, hollowing out parts of the tree for themselves. These ants use a code to stimulate the tree to produce fats and proteins that feed the ants. But this doesn't hurt the tree. In return for room and board, the ants trim away vines, which might otherwise strangle the tree, and protect the tree from other insects that would feed on it. Both the tree and the ants benefit from this relationship.

Even though scientists have not been able to figure out the code used by ants to stimulate the tree to feed them, a newly discovered species of beetle has. This beetle has been seen to evict the ants from a tree and then mimic their code so that the tree produces food. This beetle is bad news for the tree, because it does not trim away vines or protect the tree from other pests.

This astonishing relationship offers challenges to those who reject the biblical account of creation. How could ants be smart enough to give instructions to a different form of life through a code that has not even been broken by modern science? How could a tree be smart enough to understand the ant's instructions? And how could a beetle be smart enough to learn and mimic the code?

Obviously, each one of these creatures was designed with each of these special abilities by an intelligent Creator. That same Creator communicates to each of us in His Word, the Bible.

Prayer: Dear Lord, as the Word, You made all things. Then You came as our Brother to redeem us. Help me to be as filled with awe about what You teach me in Scripture as I am about the wonders of Your creation. Amen.

Bodyguard Ants

I Samuel 2:9
"He will keep the feet of his saints, and the wicked shall be silent in darkness; for by strength shall no man prevail."

Butterfly caterpillars are a prized food in nature. But the riodinid caterpillars, which metamorphose into the beautiful "metalmark" and "blue" butterflies that most of us have seen, don't have much to worry about.

While it's a jungle out there, so to speak, the riodinids live a life of relative luxury, protected by bodyguards. It's amazing enough that these bodyguards are ants. Even more amazing is that the caterpillars call the ants, apparently using their own language!

The caterpillars tap out a pulse, using special organs designed just for that purpose. These pulses closely resemble the vibratory messages used by ants to communicate with one another. The tapping brings ants running to protect the caterpillar. When there is no need for protection, the caterpillar secretes a sugary substance to feed its bodyguards.

God has invented enough wonders here to keep science busy for generations. The metamorphosis of the caterpillar into a butterfly is as wondrous as it is mystifying, even to the scientist. The intelligent communication that takes place between these very different creatures and the mutual help they provide one another is filled with complex, interrelated designs that no amount of chance could devise. And it's all for the purpose of protecting butterflies. If our Creator so cares for the butterflies, how much more does He care for you and me?

Prayer: Dear Heavenly Father, I thank You that You have invited me to come to You in prayer as a child comes to his loving father. When I fear worldly evil, help me to remember to come to You. And when I am afraid to come to You, remind me that in Jesus Christ You have forgiven and accepted me. In His Name. Amen.

The Miracle of New Life

Psalm 139:13
"For thou hast possessed my reins: thou hast covered me in my mother's womb."

How wonderful it must have been to be Adam and Eve and realize that you had been personally handmade by God Himself!

Before Darwin, people commonly believed that a miracle performed by God's own hand took place with each new child. As modern medical science learns more about the development of the unborn child, many people – including many physicians – are once again seeing the personal working of God in the developing unborn child.

When the human egg is fertilized, it is but one lone cell. That cell is uniquely and completely human, containing instructions for every detail from hair color to height. Only 10 days after fertilization the unborn child is producing hormones that communicate to the mother that her body must make changes so the infant can continue to develop. Just ten days later the nervous system, including the foundation of the brain, is already established. The next day the heart begins to beat. Only a week after this the infant already has begun to form its backbone, arms, legs, eyes and ears. Just two days after that the infant is 10,000 times larger than it was only 30 days before! In only ten more days the infant's unique brain waves can be detected and recorded. And all this in just over the first month!

Modern science understands very little about how a new human being is formed. But even if science could describe every step, each one of us would be no less than wondrously formed personally, by God's own hand.

Prayer: Dear Father, renew within me the realization that You have personally hand-made me, that You have loved and forgive me because of Your Son Jesus Christ and that You wish a closer and more intimate relationship with me. In His Name. Amen.

Microscopic Bears

Psalm 46:10
"Be still, and know that I am God: I will be exalted among the heathen, I will be exalted in the earth."

We have presented some strange and unusual creatures on previous "Creation Moments" programs. The eight-legged, egg-laying "bears" I want to tell you about today are perhaps the strangest creatures yet. They are called "bears" because they indeed look like bears – with claws, a bear-like head and the habit of climbing just like a bear. Yet these bears, often known as "water bears," are microscopic but complex multi-celled creatures.

They have a well-developed nervous system, including a brain. The females make little sacks out of shed skin in which they carry their eggs on their backs until they hatch.

Perhaps the most amazing feature of water bears is that they are incredibly over-adapted for life on Earth. They can live in fresh or salt water and have been found as deep in the oceans as 16,000 feet and as high on mountains as 22,000 feet. They can live on land, on glaciers, or in hot springs. Fifteen minutes in boiling water won't kill them, nor will being frozen of within less than a degree of absolute zero. They are unbothered by radiation levels 11,000 times higher than it would take to kill a human being.

Evolutionists say that living things develop to adapt to their environment from simpler living things. What kind of environment could allow these water bears to develop all these amazing abilities? Evolutionists admit that they are unable to account for the complexity and amazing abilities of these tiny water bears. But we are convinced that they are a tribute to the creativity of God Himself.

Prayer: Dear Lord, there is no limit to Your imagination and ability! As I deal with life, help me to be more dependent on You, knowing that as limited as I am, You are unlimited. Amen.

The Walls Did Come Tumbling Down

Joshua 6:20
"So the people shouted when the priests blew the trumpets: and it came to pass, when the people heard the sound of the trumpet, and the people shouted with a great shout, that the wall fell down flat, so that the people went up into the city, every man straight before him, and they took the city."

Because of their commitment to evolution, many archaeologists who study biblical events often declare the Bible's account of history to be in error. In the 1950s, British archaeologist Kathleen Kenyon studied the site of Jericho and declared that the walls could not have come down as related in Joshua chapter six.

Kenyon said that Joshua could not have led the Israelites against Jericho, resulting in the tumbling of the walls and the destruction of the city, because she dated the destruction of the city 150 years before Joshua could have arrived. She based her date for the destruction of Jericho on pottery fragments.

But now a Bible-believing archaeologist, Dr. Bryant Wood, has shown that the destruction of Jericho did indeed happen as the Bible relates it. He says that Kenyon misdated Jericho because she failed to find a certain type of pot. Dating soot found in building blocks from the burning of the city after the walls tumbled, Dr. Wood arrived at the correct date for Joshua's involvement. Egyptian amulets also confirmed the dating. Dr. Wood said that bushels of grain found at the site provide further evidence that Jericho was conquered rapidly.

Except for a few instances that are still being studied, every time unbelieving archaeologists have declared that the Bible is wrong, they themselves have been subsequently proven wrong. In every case, the Bible's account of history has been shown to be absolutely accurate.

Prayer: Dear Lord, as You reminded the Jews, "The Scriptures cannot be broken." I pray that You would move more Christians to confront the claims of doubters with the truth in a way which will invite those who doubt to saving faith. Amen.

The Heart Speaks to the Brain

Matthew 6:21
"For where your treasure is, there will your heart be also."

"Do what your heart tells you" is common advice. Literature, both modern and ancient – including the Bible – speaks of the heart as much more than a vital but unfeeling pump. New research has shown that these references to the heart may not be so wrong.

Researchers now describe the heart as an "intelligent" organ that regularly communicates with other parts of the body, even giving advice to the brain! It turns out that the heart manufactures a whole family of hormones that carry its messages.

Modern descriptions of the human body generally use very materialistic language to describe it. We have all heard that we are made up of nothing more than water and a couple dollars worth of common chemicals. The heart is often simply described as a pump. Human existence is reduced to nothing more than our few years on Earth. But the human body is a miraculous symphony of dependent and interrelated parts, each carefully designed with the others in mind.

Medical discoveries, such as this discovery about the heart, add weight and meaning to the Bible's description of the creation. The heart is more than a pump. Human beings are more than water and a few dollars worth of chemicals. So we can conclude that human existence is also more than the few years we spend here on Earth. And this tells us why it is important for us to learn about what God has done through Jesus Christ, so that we can have an eternal relationship with Him.

Prayer: Dear Father in Heaven, I thank You for the richness of Your love toward me. I thank You that You loved me when I did not love You and that You spared no expense, not even the life of Your only Son, so that I might receive forgiveness. Amen.

Paleontologist Ants

I Thessalonians 5:21
"Prove all things; hold fast that which is good."

Paleontology is a branch of geology dealing with the fossils of once living things. Paleontologists collect and study fossils. Before long, the paleontologist has spun a story about what the creature that once owned the bone fragment looked like and how he lived. It's not unusual for new discoveries to show that a previous paleontologist's stories were largely imagination.

But it seems that not all paleontologists are human.

Harvester ants usually collect seeds for food. But harvester ants are also found in arid deserts of the American west. They, too, collect seeds for food. But with the scarcity of plant life, harvester ants don't find enough seeds to occupy all their time. Being hard workers, they still collect. What have they chosen to collect? Strangely enough, small fossils. From the ants' perspective, some of the fossils are not all that small. One ant was seen hauling a fossilized jawbone of a small rat-sized marsupial to its nest. Fossils are deposited on top of the ant's mound. Some mounds may have hundreds of fossils on them.

Harvester ants bring their fossils from as deep as six feet below the surface, and human paleontologists have learned that, by studying these ant collections, they can gain some idea of the larger fossils they may find below. But why would harvester ants collect fossils in the first place? After all, they have been at it for far longer than we humans have. Could it be that the ants know something we don't?

Prayer: Dear Heavenly Father, grant me wisdom and understanding so that I may have a quick mind, guided by Your Holy Spirit – a mind which weighs everything I see and hear against the truth of Your Word. In Jesus' Name. Amen.

Ref: Adams, Daniel. 1983. "Science in pictures." *Science 83*, July/August.

Literate Butterflies

Psalms 96:6
"Honour and majesty are before him: strength and beauty are in his sanctuary."

Did you know that you could spell the word "butterfly" with butterflies? Or you could spell your own name with butterflies. You could even write the year of your birth with butterflies. And if you had enough butterflies, you could even include the month and day you were born.

Nature photographer Kjell Sandved has assembled one of the most unusual and amazing butterfly and moth collections yet. Traveling all over the world, Sandved has photographed the wings of many of the world's 200,000 species of butterflies and moths. This project began more than 30 years ago when he worked at the Smithsonian's museum of Natural History. One day as he was looking through uncatalogued specimens, he saw a butterfly wing with a perfect representation of the letter "F."

Since then, Sandved has photographed several representations of every letter of the English alphabet, all the arabic numerals and many non-English letters on the wings of butterflies and moths. In addition, he has found many images of plants and animals as well as human faces on the wings of these delicate insects.

Many of the markings on butterflies and moths are part of their camouflage. In other cases, like the giant eye spot on each wing of some butterflies and moths, the spot is known to keep birds away. Beauty and protection with a little bit of wonder thrown in – surely God has done all things exceedingly well!

Prayer: Dear Lord, whether I look at a sunset, a butterfly or Your plan of salvation, I am filled with wonder at the completeness of Your wisdom and Your generosity. Make me a better witness to others about what You have done. Amen.

Ref: Amato, Ivan. 1990. "Insect inscriptions." *Science News*, v. 137, June 16. p. 376.

The Vanishing River

Genesis 7:11
"In the six hundredth year of Noah's life, in the second month, the seventeenth day of the month, the same day were all the fountains of the great deep broken up, and the windows of heaven were opened."

In Italy the river is called the Timavo River, and some of it is missing. The river actually begins in Yugoslavia, finally reaching the Adriatic Sea through Italy. Now it would be strange enough if the beginning of the river were missing, but in this case it's the middle of the river that's missing.

Two thousand years ago, the poet Virgil made reference to the strange fact that the river disappears into the ground. He wrote that he thought the river actually led to the gates of Hades. Not until the 1920s did scientists prove that the river resurfaces only a mile from the Adriatic Sea near the Italian town of Duino.

Why does the river disappear? It flows on sandstone over porous Karst limestone, which is very much like Swiss cheese. When the river wears away the sandstone, it begins to seep into the holes in the limestone. As it does so, it dissolves the limestone, cutting an underground bed for itself. In some places the river is nearly 1,000 feet underground.

Where the Timavo River flows out of the ground toward the sea, we have a miniature and tranquil picture of the great water resources within the Earth that were released when the fountains of the deep were broken up at the beginning of the Genesis Flood. While a worldwide flood may seem too unusual for modern geology to believe, stranger phenomena are known to geology – like the Timavo River!

Prayer: Dear Lord, if You limited Yourself to working in ways we could figure out ourselves, You would be as limited as we are. Let this fact inform my faith when I have difficult-to-understand passages in Your Word which many in the world refuse to believe. Amen.

Ref: Hansen, James. 1984. "The river vanishes." *Science 84.* p. 84.

Georgia's Grand Canyon

II Peter 3:5-6
"For this they willingly are ignorant of, that by the word of God the heavens were of old, and the earth standing out of the water and in the water..."

Providence Canyon, near the town of Lumpkin in western Georgia, looks so much like Arizona's Grand Canyon that it is often called the Little Grand Canyon. The walls and peaks of the canyon feature striking layered shades of tan, buff, pink, salmon, lavender and deep orange-red.

The Little Grand Canyon is really nine finger-like canyons that are nearly 200 feet deep. The canyon has been made part of a 1,100-acre park. Visitors to the park may tour the canyon itself. But the most amazing fact about the Little Grand Canyon is that it is only about 150 years old!

This area of western Georgia is made up of loosely arranged sand and clay sediments. The entire area was once under water, and many of the sediments never hardened into sedimentary rock. About 150 years ago, the forest was cleared from the area, leaving the ground exposed to the forces of erosion. Heavy rains have deepened the canyon by as much as six feet in one night.

Many geologists who believe in the Creator have studied the Grand Canyon in western North America. They generally agree that this canyon was formed in much the same way as Georgia's Little Grand Canyon has been formed. As draining waters of the great Flood cut through the still soft sedimentary layers laid down by the floodwaters, the major portion of the Grand Canyon as we know it today was formed. Georgia's Little Grand Canyon is a miniature laboratory that shows us how the great Grand Canyon came to be.

Prayer: Dear Heavenly Father, Noah lived in an evil and unbelieving age which was even worse that ours today. I thank You for fellow believers. Help me to provide Christian encouragement to others, and use fellow Christians to encourage me. In Jesus' Name. Amen.

Ref: Dickenman, John. 1988. "Georgia's Little Grand Canyon." *Georgia Journal*, Spring. p. 48.

The Body's Amazing Ability to Adapt

Genesis 1:28
"And God blessed them, and God said unto them, Be fruitful, and multiply, and replenish the earth, and subdue it: and have dominion over the fish of the sea, and over the fowl of the air, and over every living thing that moveth upon the earth."

God has given our bodies the ability to adapt to an amazing range of different circumstances. We're all familiar with changes that take place in our bodies when we gain weight or when we take up exercise to lose weight and become more fit. But these changes, great as they can be, are only the tip of the iceberg.

A study released some time ago showed that when people in their 70s and 80s began an exercise program, not only did their muscles become stronger, so did their bones! Some in the program were even able to give up their canes. But our activities change our bodies in ways you might not otherwise expect.

Perhaps one of the most incredible adaptations the human body is capable of is found among the Quechua Indians high in the Andes. Living at 12,000 feet, their bodies need to be much more efficient than most in order to collect enough oxygen and circulate it in their bloodstreams. For this reason, their lungs are able to oxygenate 2% more blood than those who live in lower altitudes. In addition, they have unusually large hearts to circulate the blood. Because the mountainous climate is cold, they have short bodies that tend to conserve heat. Their hands and feet have extra blood vessels for more efficient circulation. The circulation to their feet is so efficient that they can walk barefoot on ice and snow and never get frostbite.

When God told Adam and Eve to fill the Earth, He had already given them the ability to adapt to a wide range of conditions.

Prayer: Dear Father, I thank You that You have given me this wonderful body with all its abilities. Although because of sin my body is not perfect, I give You thanks for its abilities and ask that You would grant me, according to Your will, good health. In Jesus' Name. Amen.

Deceitful Orchids

Isaiah 45:22
"Look unto me, and be ye saved, all the ends of the earth: for I am God, and there is none else."

Orchids fool bees and other insects, using a number of amazing methods to get insects to carry their pollen to other flowers of their kind.

The red helleborine orchid uses such a trick to get bees to carry its pollen to other helleborine. The orchid has no nectar to attract the bee and make a stop worth its while. The bellflower, which is not an orchid at all, does have nectar that is favored by bees. So the helleborine orchid mimics the bellflower in shape and even in color. The red shading of the orchid is so slightly different from the bellflower that bees cannot see the difference. By the time an unsuspecting bee has completely searched the orchid for nectar, the bee is well covered with pollen. It then flies off looking for a bellflower, and is just as likely to find a female orchid waiting for the pollen he carries.

Evolutionists say the helleborine orchid developed this strategy over millions of years so that it, too, could be pollinated. But think about that for a moment. How could an orchid – which cannot see – decide with a brain it doesn't have to make itself look like another flower? If orchids were that smart and had the power to change themselves, it would have been a lot easier for the orchid to simply begin producing its own nectar to attract bees!

This orchid is really just one more evidence that God has provided to help us see, in the complex interrelationships of unlike creatures, that there is one all-wise and all-powerful Designer and God over all creation.

Prayer: Lord, I know that You have made me and that on my own I cannot change myself for the better. Help me to be more dependent on You and trusting of Your Word to me in Scripture so that I might walk in closer fellowship with You. Amen.

The World's Most Amazing Bears

Job 38:3-4
"Gird up now thy loins like a man; for I will demand of thee, and answer thou me. Where wast thou when I laid the foundations of the earth? declare, if thou hast understanding."

The little eight-legged "bears" known as water bears offer such a challenge to evolution that high school students are often encouraged to skip over the textbook section dealing with them.

One of the most amazing abilities these microscopic water bears have is the ability to change their form in order to survive difficult conditions. In an unfavorable environment, the bears will roll in their heads and legs until they look like tiny barrels. During this transformation, their tissues lose most of their water, and the tiny barrel that was once a "bear" is now no larger than one of its legs before the transformation.

In this condition, the bear can survive almost any conditions. They have been frozen to within less than one degree of absolute zero. They have been held in liquid air for up to 20 months and boiled for up to 15 minutes. They have been subjected to poisonous atmospheres and exposed for 24 hours to 11,000 times the radiation it would take to kill a human being. But as soon as conditions became favorable again, the nearly indestructible "bears" absorbed water and were soon climbing around, as healthy as could be.

It's obvious why many who believe in evolution skip over the description of water bears. Evolution teaches that our abilities and features evolved in response to our need to survive. Evolutionists say that what we are was not given to us by God. But it is clear that water bears never had to evolve to survive under these extreme conditions. Their remarkable abilities are obviously the gift of our Creator!

Prayer: I thank You, dear Lord, for creating "water bears" with such amazing abilities that they make those who deny Your creating hand uncomfortable. Use me as Your witness among men so that they may be led to Your forgiving love in Jesus Christ. Amen.

Ref: Vetter, Joachim. 1990. "The little bears that evolutionary theory can't bear!" *Creation Ex Nihilo*, v. 12, n. 2, Mar.-May. p. 16.

Fish Teach Humans to Make Better Ceramics

Genesis 1:31
"And God saw every thing that he had made, and, behold, it was very good. And the evening and the morning were the sixth day."

When scientists finally learn how to make ceramics that can endure high temperatures and conduct electricity without resistance, they may have to thank the sea urchin for teaching them how to do it.

While the ceramics that humans manufacture are fairly strong and resist forces that destroy other materials, they have their imperfections. They are not shatter resistant. And it takes a lot of heat to create them. On the other hand, mollusks like the nautilus and sea urchin make shatterproof ceramics out of calcium carbonate – which is chalk – using no heat and a little water. And the mollusk-created ceramics come in intricate shapes often much more complex than those made by humans.

Scientists are now studying how mollusks make their ceramics so that we can also make better ones. The processes they are learning will enable the manufacture of strong ceramic materials that conduct electricity without resistance. They will be cheap and easy to make, yet they will provide us with better building materials and even artificial bones.

Scientists are learning that the secret to making superior ceramics uses cheap materials and a very complex series of chemical reactions carried out by special cells in ceramic-making mollusks. It's definitely not a system that was worked out by no one at all through chance and accident. In effect, science is seeking to learn how the Creator made ceramics, so that we can do it too!

Prayer: Father, I often forget that Your wisdom extends to very material things, things which I don't usually associate with the spiritual. Teach me not to separate the spiritual and material, but see them both as coming from Your Hand. Help me to glorify You in spiritual as well as material matters. In Jesus' Name. Amen.

Ref: Amato, Ivan. 1990. "Better ceramics through biology." *Science News,* May 5. p. 287.

The Origin of Modern Language

Genesis 11:9
"Therefore is the name of it called Babel; because the LORD did there confound the language of all the earth: and from thence did the LORD scatter them abroad upon the face of all the earth."

Just where on earth did our languages come from? Those who believe that the Bible's account of human history is reliable would expect the answer to pinpoint a general area of Asia near where the Ark of Noah landed.

Of course, most historians don't take the Bible seriously. They believe languages simply evolved from grunts and growls. They would expect languages to trace back to the general areas in which those languages are spoken today.

An article in the March 1990 *Scientific American* reports on a study that tried to find out the origin of various European languages. The study used many pieces of information, including similarities between various European languages, to determine where the original language they all came from had its start. Earlier investigators had said that European languages developed in Europe. But this later study concluded that European languages actually had their start in a general area of eastern Anatolia, near Ararat.

This study, which impressed other researchers because of its thoroughness, means that historians can no longer ignore the Bible's account of history. And we expect that additional research that has been suggested to confirm these findings – as well as research into the origins of other language groups – will provide even more powerful, hard-to-ignore evidence for the Bible's presentation of history.

Prayer: I thank You, Lord, for the wonderful gift of language. And I give You special thanks for communicating to us in Your Word. Help me to place a higher value on what You say to me in Your Word so that I am moved to learn more of it. Amen.

Ref: "European languages origin near Ararat." *ARCHAEOLOGY and Biblical Research*, v. 3, n. 2, Spring 1990. p. 60.

Scientists Learn from Flies

Proverbs 1:29-31
"For that they hated knowledge, and did not choose the fear of the LORD: They would none of my counsel: they despised all my reproof. Therefore they shall eat the fruit of their own way, and be filled with their own devices."

Your eye is a miraculous system of individual parts, all of which work together to provide vision. Light must be collected, focused, interpreted and then translated into signals that are sent to the brain, where the information is further processed and turned into the images we see. For as much as we know, modern science still isn't quite sure how the whole system works.

Scientists would like to learn just how vision works so they can build robots that can see. Some experimental robots have been built that can see after a fashion, but even the common housefly can see better than the best machines made by science. Scientists are now studying the fly's eye in order to learn more about how vision works.

The fly's eye, with its 3,000 lenses, is remarkably complex and similar to the human eye. There's no sign of evolution here. Even the fly's nervous system, which is essential in translating sight, is highly complex. Each of the fly's 3,000 lenses collects light into a tubular structure. This structure contains eight light-sensitive cells. These cells process the light and turn it into electrical signals that then go to the fly's brain.

It's easy for people to say that there is no Creator. But as soon as they begin to study any living system, they are quickly humbled and eager to learn from it. That in itself is an excellent argument for the existence of the Creator.

Prayer: Dear Father; I thank You that even when man denies You, You have the last word. Help me to see those actions in my life which deny You as Creator and Jesus Christ as my Savior and with Your help make a better confession. In His Name. Amen.

Ref: Spice, Byron. 1990. "Scientific vision." *Albuquerque Journal*, Apr. 23. p. 8, Sec. B.

Did the Sun Really Stand Still?

Joshua 10:13
"And the sun stood still, and the moon stayed, until the people had avenged themselves upon their enemies. Is not this written in the Book of Jasher? So the sun stood still in the midst of heaven, and hasted not to go down about a whole day."

In the tenth chapter of the book of Joshua we read how, as Israel fought for occupation of the Promised Land at Gibeon, Joshua asked the Lord to stop the sun so that Israel could finish its battle. According to Joshua 10:13, the sun stopped in mid-sky and did not set for nearly an entire day. According to the most accurate dating, this was about 1400 BC.

Our modern scientific age knows that the Earth, which weighs billions of tons, revolves around the sun while turning on its axis. In order for the sun to stand still in its apparent movement across the sky, the Earth would have to stop spinning. The Earth spins at a speed of about 1,000 miles per hour. If it were to suddenly stop, the force would rip the Earth apart. All of this is what science tells us. And from these facts, science therefore concludes that the sun could not have stopped in its path across the sky. But this final conclusion is not scientific, even if all the other facts are true.

Joshua 10 tells us that God made the sun stop in its apparent path. Science admits that it cannot tell us what God can or cannot do. Science can only tell us that it would be very unlikely for the sun to stand still in the sky. But Joshua 10:14 tells us this and more – this had never happened before and had not happened again at the time Joshua was written.

It's not surprising, then, to learn that, according to Aztec lore from Mexico, virtually on the other side of the world, the sun failed to rise for an entire day in the City of the Gods around 1400 BC.

> **Prayer: Dear Lord, our age is a materialistic age where we deny what we cannot understand. I confess that I cannot understand how the sun could appear to stop in the sky, but I know that Your Word says that You did this great miracle. Grant me your Holy Spirit so that I may not be infected with the doubt that so permeates the world. Amen.**

Ref: Sitchin, Zecharia. 1990. "How Jericho fell." *The Christian News*, Apr. 30. p. 17.

The Bible Speaks About Our Ecological Crisis

Hosea 4:2-3
"By swearing, and lying, and killing, and stealing, and committing adultery, they break out, and blood toucheth blood. Therefore shall the land mourn, and every one that dwelleth therein shall languish, with the beasts of the field, and the fowls of heaven; yea, the fishes of the sea also shall be taken away."

We hear so much about the ecological crisis today. We hear that we'll be buried in our own garbage or that there are too many people for the Earth to support and that lakes, rivers and forests are all dying.

The world tells us that the cause of the ecological crisis is too much garbage, too much manufacturing, too many people and too much greed. According to some people, recycling, less waste of resources, and abortion are important in solving the problem. Of course, killing the next generation through abortion is not an answer. While being less wasteful is always good advice, even if we stopped all the factories in the world and recycled everything, these solutions don't get to the real problem. The Bible tells us the real problem.

According to Hosea 4:2-3, a land in anguish and a loss of birds, animals and fish are not the results of 20th-century technology and population. These things have all happened before modern technology came on the scene. According to God, the cause of these ecological disasters is people and what they do. It is swearing, deception, murder, stealing, violence and bloodshed that cause this death to the land.

God's Word also offers us the solution. That solution is found in the forgiveness of sins and new life that is ours in Jesus Christ. Romans 8:19-22 tells us how this influences the whole creation.

Prayer: Dear Father, the greed and violence of this present world has influenced me, too. I confess that my love does tend to grow cold in our violent world. Once again, give Your people boldness to witness Christ rather than the fear which we find in the world around us. In Jesus' Name. Amen.

The Games Animals Play

Job 40:20
"Surely the mountains bring him forth food, where all the beasts of the field play."

What is play good for? Well, it's lots of fun, but fun has no evolutionary value. And evolutionists are puzzled over the fact that animals seem to love to play. They have offered many theories about how play might have some survival value, but they haven't convinced even each other that any of their explanations for play really explain anything.

Most of us have seen puppies and kittens play, sometimes even playing with each other. A common sight on a farm is a lone colt, leaping and dancing in the field for no apparent reason other than the joy of doing it. Many creatures enjoy a good old-fashioned game of king of the mountain.

Many of the games animals play show a surprising amount of creativity. Right whales have been seen to raise their tails into the air at right angles to the wind, using them as sails to propel themselves toward shore. Polar bears enjoy dropping objects down slopes and then chasing them. Some cubs even play throw and fetch games with themselves. Monkeys often make sponges out of chewed leaves in order to get drinking water out of tree hollows. But they often get bored with this and start clowning, starting water fights and making faces with the leaves in their mouths.

According to the Bible, our Creator is capable of joy. It would seem that He has given many of His creatures the ability to enjoy life and have fun too. Play is a gift of our Creator that makes life just a little richer.

Prayer: Lord, I thank You for the ability to enjoy Your good gifts and have fun. I ask that these things would have a proper place in my life, making it richer, and filling me with more gratitude toward You. Amen.

REF: Fagen, Robert. 1983. "Horseplay and monkeyshines." *Science 83*, Dec. p. 71.

Nature's Chemical Wars

Job 10:10-12
"Hast thou not poured me out as milk, and curdled me like cheese? Thou hast clothed me with skin and flesh, and fenced me with bones and sinews. Thou hast granted me life and favour, and thy visitation hath preserved my spirit."

It's a jungle of chemical warfare out there, and some of it is having an effect on human health.

Scientists are learning that plants, including those we eat, are engaged in chemical warfare that dwarfs even the human use of pesticides. Researchers have learned that many plants produce powerful pesticides in order to protect themselves from their enemies.

Flies are a big problem for citrus fruits like oranges. Research has shown that peels from oranges and other citrus fruits produce a variety of oils that serve as powerful insecticides. In one experiment, nine houseflies confined to a chamber with orange peels were out cold within 15 minutes. After two hours they were dead. Mushrooms, celery, figs, potatoes and herbs also produce powerful insecticides, and some of them are ingested at levels thousands of times higher than pesticides produced by humans. Some of these natural pesticides are being investigated as possible causes of cancer. That so-called health food may not be so pesticide free after all.

Life itself is built on very complex and sophisticated chemistry. Many scientists who once accepted evolution have abandoned that theory because they know this chemistry is not accidental. Now we see that even plants generate their own pesticides to kill insects that are harmful to them. This speaks even more eloquently of a loving Creator who cares for His entire creation.

Prayer: Dear Father, I thank You that You care about the creation You have made. Help me to care about it, too – if for no other reason than that You care about it. And never let me forget that You care about me and have provided me with a relationship with You through Your Son, Jesus Christ. In His Name. Amen.

Hibernation: Not Simply Sleep

Psalm 121:2-4
"My help cometh from the LORD, which made heaven and earth. He will not suffer thy foot to be moved: he that keepeth thee will not slumber. Behold, he that keepeth Israel shall neither slumber nor sleep."

As winter settles over the northern hemisphere of our planet, many kinds of creatures are moving toward hibernation. True hibernation is much more than just a deep, prolonged sleep. In fact, many animals that are popularly believed to hibernate, like bears, are not true hibernators. A wintering bear's body temperature seldom falls below 86 degrees, so bears are easily awakened. True hibernators actually need several hours to awaken.

The ground squirrel is a true hibernator. Its hibernation pattern is triggered by an internal clock that causes hormone changes. These changes not only lower the squirrel's temperature, metabolic breathing and heart rate, but also changes the way its nervous system and cell membranes operate. If the squirrel's nervous system and cell membrane operation were not modified for hibernation, the other changes would kill it.

Once the squirrel is in true hibernation, its body temperature drops to about 35° F, its heart rate drops from 350 beats per minute to about 3 beats per minute, and it will breathe only once every several minutes.

The vast and complex internal changes that must take place in hibernation, affecting the function of every cell, show us that hibernation is an ability that was built into many creatures. If all these changes were due to genetic accidents, there would be no hibernating animals today. They would all have died trying to find the right combination of internal changes to allow hibernation to occur.

Prayer: Dear Heavenly Father, I see Your love for all that You have made in the ways that You have provided for the good of all Your creatures. If You would be their Father through such tender loving care, how much more would You be my Father for the sake of Your Son, Jesus Christ! Amen.

Ref: Fleming, Carol B. 1984. "How do animals hibernate?" *Science 84*, p. 28.

New Discoveries About that Terrible Lizard

Job 40:7-8
"Gird up thy loins now like a man: I will demand of thee, and declare thou unto me. Wilt thou also disannul my judgment? wilt thou condemn me, that thou mayest be righteous?"

Throughout the 20th century, textbooks and movies have depicted the most terrifying of all the dinosaurs. Scientists reconstructing Tyrannosaurus Rex from fossilized bones of the monster say he stood 20 to 30 feet tall and was 40 to 50 feet long. This meat eater had six-inch-long, dagger-like teeth and eight-inch-long talons. No wonder *Tyrannosaurus* has been the monster of choice for so many films.

You might be surprised to learn, then, that only seven fossilized Tyrannosaurs skeletons have ever been found, and none of them was complete. In fact, it was not until 1990 that scientists found a complete set of arm bones for the creature.

The discovery of this first complete set of arm bones will help settle a debate among scientists over whether Tyrannosaurs could actually use those arms to help capture and eat his prey. And only within the last few years have scientists changed their reconstruction of Tyrannosaurs from a tail-dragging creature to one that held its powerful tail in the air.

It's clear that if scientists have had to change their conclusions about something as basic as the structure of Tyrannosaurs, there are many other facts about him of which they can't be sure. Did humans live at the same time as Tyrannosaurs? Most scientists say no. But there's no evidence that actually proves that. And some day evidence may be uncovered that validates the Bible's history, placing humans and dinosaurs together on the early Earth.

> *Prayer: Dear Father, how wondrous Your creation must have been before it was destroyed by sin! Do not let sin destroy me, but fill me with the spirit of a pilgrim who longs to be home in the new heavens and earth which are coming. In Jesus' Name. Amen.*

Ref: Morrison, Patt. 1990. "Montana plans to keep its Tyrannosaurus Rex skeleton." *Minneapolis Star Tribune*, Aug. 12. p. 21A.

How to Speak Crow

John 1:14
"And the Word was made flesh, and dwelt among us, (and we beheld his glory, the glory as of the only begotten of the Father,) full of grace and truth."

While many cities around the world have had troubles with pigeons, in recent years the Kremlin has had problems with crows. It seems that the Kremlin crows were amusing themselves by skating on their claws down the onion-shaped domes. The problem was that their claws were removing ribbons of gold leaf that covers the outside of the domes. To solve their problems, officials turned to science.

Soviet scientists have compiled glossaries of crow calls and their translations, as well as glossaries of many other bird calls. Scientists knew that crows were too smart to fall for a recorded crow voice sounding an alarm – at least for very long. Since scientists knew the meanings of several of the crows' calls, they decided to broadcast a recording of the crow call that translates, "I've been caught by a falcon. Please mob it until it lets me go."

As expected, the crows mobbed the loudspeaker, trying to free their comrade. At the same time, a falcon was released from the ground who then rose above the crows to dive at them from above. Unable to come to terms with an invisible lying comrade, the crows have for the most part left the Kremlin alone.

While humans clearly have the most complex languages of any creature, it is also clear that the traditional evolutionary story of how human language evolved makes no sense. The Word Himself who created us and all creatures is clearly the author of all language and intelligent communication.

Prayer: Dear Lord, I thank You for the gift of communication, and especially for Your communication to me and all mankind in the Bible. Let me not misuse Your gift of communication either by neglecting Your Word or by speaking that which is not true. Amen.

Ref: Boswall, Jeffery. 1987. "Russia is for the birds." *Discover*, Mar. p. 78.

Instant Fish

Psalm 100:1-3
"Make a joyful noise unto the LORD, all ye lands. Serve the LORD with gladness: come before his presence with singing. Know ye that the LORD he is God: it is he that hath made us, and not we ourselves; we are his people, and the sheep of his pasture."

Imagine that you want to catch some fish, but all you see before you is dry sand. It hasn't rained in some time, and there is no sign of life. Now imagine that you also have a large tanker truck of water. So you pump the water out of the tanker and let it run into a depression in the sand. Let's say it takes you ten minutes to shut off the water and pull a net from the front seat. By the time you get to the small pool you've created, you can net fish.

That's exactly what researchers did. And they caught instant fish! These fish are called salamander fish. They live only in the on-again, off-again lakes and ponds of southwestern Australia. When there is plenty of rain, the fish live in their ponds, feeding on insect larvae. But when the ponds dry up, the fish follow the water table underground, burrowing into the sand and entering a kind of dormant state, apparently breathing through their skin.

The salamander fish has no living relatives and doesn't seem to be related to anything else in an evolutionary sense. Its skull is extra large and strong for its size and its spinal bones are separated, providing extra power and mobility for digging in the sand.

While millions of lakes and ponds with fish have dried up throughout history, there is no evidence that fish have ever learned to change their way of life so that they could continue to live during dry spells. The salamander fish was obviously created with this very special ability. Just as the Bible says, all creatures were created fully formed by God.

Prayer: Dear Lord, the salamander fish has no one but You to thank for its wonderful abilities. I thank you that this is true for all of us, and I ask that You would help me learn to see and use all the abilities You have given me to Your glory. Amen.

Ref: Benowitz, Steve. 1990. "Of instant fish and pickled sharks." *Ohio State Quest*, Summer. p. 15.

Plastic Ceramics

Psalm 10:4
"The wicked, through the pride of his countenance, will not seek after God: God is not in all his thoughts."

All of us are familiar with ceramic materials. Humans have created ceramic materials for thousands of years to make pots, dishes and other containers. Ceramics endure high levels of heat and they are tough and hard. Utility companies have used ceramic insulators on their poles for generations because ceramics are good electrical insulators.

But ceramic materials have had secrets they are only now giving up. In 1986 it was discovered that certain types of ceramic materials could be made that are the opposite of good insulators. Not only do they conduct electricity, but at low temperatures they conduct electricity perfectly, making them superconductors. No one ever expected ceramics to do that.

New ceramic materials have been created that do something else no one ever expected – they can stretch! Researchers have reported creating ceramic strips that can be stretched, while still hot, to 2.5 times their starting length. This means that it may be possible to mold ceramics into complicated shapes, like engine parts that will resist wear much better than current materials.

God had absolutely no limitations when He designed each of the materials that make up our visible world. Science done properly is, as Sir Isaac Newton said, the attempt to think God's thoughts after Him. In other words, as we learn about the universe, we need to set aside our limited ideas in order to discover things we could never imagine – but God could. That's why those who believe in the Creator can be better scientists than those who don't.

Prayer: Dear Heavenly Father; compared to Your thoughts, my thoughts are nothing. Remove human pride from my thinking and fill me with faith and understanding which is informed by Your Word so that my thoughts may be more like Yours. In Jesus' Name. Amen.

Ref: "Ceramics go to new lengths." *Science News*, Mar. 31, 1990. p. 199.

One Smart Woodpecker

Acts 17:28
"For in him we live, and move, and have our being; as certain also of your own poets have said, For we are also his offspring."

While most woodpeckers carve out cavities in dead trees for their nests, the red-cockaded woodpecker prefers to use the living pine trees of the southern United States. The woodpecker always chooses one of these living southern pines that is over 60 years old. While these pines can live 200 years or more, usually at about age 60 a fungus infection begins to rot away the heart wood at the center of the tree.

Over months or even sometimes years, the woodpecker works his way through the living tissue to the hollow center where he will build his nest. The nest cavity is always above the level of shrub growth in order to provide a safe haven from any forest fires.

In addition, surrounding foliage can give the gray rat snake access to the nest. These snakes, natural enemies of the woodpecker, can also climb pine trees. But the snake does not like the gummy resin of the pine tree. In fact, snakes that encounter the resin will begin to writhe and fall to the ground. So to make sure that its home is absolutely safe, the woodpecker drills a series of holes around the hole leading to its nest. The holes are kept open so that the nest hole is always protected by the resin.

The red-cockaded woodpecker carefully selects a site for its home in order to be safe from both fire and a natural predator – using the natural chemistry of the tree to aid its survival. Obviously this intelligence was given to the woodpecker by the Creator since trial and error cannot explain such careful planning.

> *Prayer: Dear Father in heaven, the entire creation bears witness to the fact that You are a caring God not distantly removed from us. Grant me a clearer sense of Your presence in my life and Your desire to be even closer to me than You are now. In Jesus' Name. Amen.*

Ref: Mohlenbrock, Robert H. 1990. Bienville Pines, Mississippi. *Natural History,* Aug. p. 29.

Astonishing "Coincidences"

Psalm 40:5
"Many, O LORD my God, are thy wonderful works which thou hast done, and thy thoughts which are to us-ward: they cannot be reckoned up in order unto thee: if I would declare and speak of them, they are more than can be numbered."

A coincidence occurs when two events that you know are not related happen in such a way that they seem related. Let's say that you receive an unexpected bill in the mail for $124.00. And in the same day's mail you find a check for $124.00 with a note from a friend saying, "Here's the $100.00 you loaned me four years ago with some interest." What a happy coincidence.

When more than two seemingly unrelated events seem to work together, you would have to conclude that they were, indeed, related.

That's why scientists are beginning to wonder whether there isn't a plan to the universe after all. For life to exist in the universe, literally millions of details must come together in just the right way. If the force of gravity or the forces holding the atom together were only a tiny bit stronger or weaker, life could not exist. The carbon atom is the basis of life. Is it coincidence that the carbon atom is unique in its ability to form the long, complex chains of molecules needed for life? If stars burned at a slightly different rate than they do, life would either be burned up or frozen to death.

Many scientists have actually stated in writing that they are astonished at the huge number of cosmic "coincidences" that have come together in the universe to make life possible. Some are even suggesting that the universe has an impersonal intelligence. But why cloud the issue this way? Why not simply recognize that there is indeed a Creator God who has created a whole universe just for the benefit of life – especially humankind?

Prayer: Lord, I thank You that despite man's studied unbelief, You have made Yourself and Your concern for mankind so clear that even unbelievers must recognize it. Use me and all of Your people to clearly voice the more complete details of Your love for all mankind in Your innocent suffering and death on our behalf. Amen.

Ref: Mallove, Eugene F. 1985. "Scientists puzzle over coincidences of the cosmos." *Minneapolis Star and Tribune,* Nov. 4. p. 11A.

The Miracle Mother

Psalm 27:10
"When my father and my mother forsake me, then the LORD will take me up."

Marsupials like kangaroos bear their young alive. However, newborn kangaroos are very tiny and immature. After it is born, a young kangaroo lives in its mother's pouch, nursing within the pouch itself. Eventually the young joey is large enough to venture from the pouch. As he grows, his time outside the pouch increases until the mother refuses to let him back in.

But things can get complicated. For example, what happens if there is still a young joey suckling in the pouch when another is on the way? If this happens, the fertilized egg does something unknown to other mammals. It will mature to the blastocyst stage and then become dormant, and it will remain dormant for up to 200 days while the joey already in the pouch grows.

The mother kangaroo is able to nurse two joeys of different ages and milk-needs in her pouch. When the second, much younger joey finally arrives in her pouch, a second kind of milk formula is ready and waiting. So both joeys can suckle, each receiving a different milk formula ideal for its age! And at the same time, yet a third offspring, dormant in the blastocyst stage, can be waiting for a vacancy in the pouch.

These astonishing abilities truly make the kangaroo a miracle mother. The Creator has designed an extremely complex arrangement to ensure that His creation, the kangaroo, continues to exist. If He has so carefully made such detailed provision for the kangaroo, how much must He care about you and me.

Prayer: Dear Lord, as Creator of all, You are also the designer and Creator of motherhood. I know from this that You understand tender loving care. Let me be reminded of this, especially when my life is filled with difficulties and frustrations, so that I might be drawn more securely to You. Amen.

Is the Bible a Book of Medicine?

Numbers 19:18
"And a clean person shall take hyssop, and dip it in the water, sprinkle it upon the tent, and upon all the vessels, and upon the persons that were there, and upon him that touched a bone, or one slain, or one dead, or a grave:"

"The Bible is not a book of science." We hear and read that statement a lot these days. It's true that a surgeon will not be found peering into a Bible during surgery in order to perfect a surgical technique. But there is a deceptive lie hidden in the claim that the Bible is not a book of science.

Back when Ignaz Semmelweis was a doctor, one out of every six women who gave birth in his hospital died of what was called "child bed fever." Dr. Semmelweis set out to discover why this was happening. He discovered that doctors were not cleaning their hands or instruments between patients. As a result, they were spreading germs from one patient to the next. Dr. Semmelweis instituted a policy requiring clean hands and instruments for each patient, and mortality rates dropped almost to zero immediately!

But Dr. Semmelweis's discovery was not new knowledge. Thousands of years earlier, God had taught the Israelites, through Moses, that whenever they came into contact with a dead or diseased person, they were "unclean." Unclean people and their clothing had to be cleansed in clear running water. They also had to sprinkle their clothing with wet hyssop branches. Today we know that hyssop contains a powerful antibacterial and antifungal agent.

So when the Bible says something that touches upon an area of science, it is still technically accurate and correct. After all, the Author of the Bible is also the Author of everything that science studies!

Prayer: Dear Heavenly Father, the world declares the Bible, Your Word, out of date and then ignores it, to its own great loss. I, therefore, ask you to forgive me for Jesus' sake, for my own neglect to make Your Word more a part of my life. Help me to abandon the world's way and make the Bible a practical part of my everyday life. Amen.

Ref: Thompson, Bert. 1990. "Dr. Semmelweis & the Bible." *Reasoning from Revelation,* June. p. 3.

The Power of God's Power

Job 26:13-14
"By his spirit he hath garnished the heavens; his hand hath formed the crooked serpent. Lo, these are parts of his ways: but how little a portion is heard of him? but the thunder of his power who can understand?"

Modern science generally holds that the sun creates its energy by nuclear fusion. However, the evidence doesn't support this theory because one product of fusion, called a neutrino, should be pouring out of the sun at a far greater rate than has actually been found. But nuclear fusion is still favored by most scientists because it is the only process they know of that can allow the sun to turn out its energy for the billions of years they think the universe has existed.

If the sun does indeed create energy through some sort of nuclear fusion, a number of different nuclear reactions could be taking place. The most common reaction would be the fusion of two hydrogen nuclei into one helium nuclei. Each time this takes place, one of the results is 25 million electron volts of energy. By comparison, the human eye needs only about 14 electron volts of energy to see an image on a dark night. But on the sun, 600 million tons of solar hydrogen are consumed this way every second! No wonder we dare not look at the sun! What's more, our sun plus billions of others have been turning out this energy for thousands of years.

We have no way of comprehending the energies God set into motion when He created the universe! And even modern science is stumped. As Princeton professor John Bahcall wrote in *Scientific American,* "Some flaw seems to exist either in the current models of the sun or in our present understandings of the laws of physics."

Even the best minds of modern science are nothing compared to the Creator's power and wisdom!

Prayer: Dear Father, no one can stand before Your power and wisdom! Fill our age with a better sense of what it means to be creatures of a powerful, yet loving God, Who spared not even His own Son for mankind's salvation. Amen.

Ref: Bahcall, John N. 1990. "The solar-neutrino problem." *Scientific American*, May. p. 54.

Harvesting Termites

Proverbs 6:6-8
"Go to the ant, thou sluggard; consider her ways, and be wise: Which, having no guide, overseer, or ruler, Provideth her meat in the summer, and gathereth her food in the harvest."

While most termites look for food underground or in mud tunnels they build over dead wood, some are literally farmers, cutting and storing hay.

Many termites will go about in the dark of night chewing off pieces of wood that are taken back to their nest. In the nest, the wood is stored or cured before being eaten. Some termites cannot digest the wood, so they keep underground gardens of fungi that break the wood down for them so they can digest it. But in any case, most termites never see the light of day – or even the darkness of night.

But there is one species of African termite that seeks its food in the open. This termite lives in large, dome-shaped mounds that cover a huge radiating network of tunnels. On warm nights, the termites leave their nest and go out onto the open grasslands looking for dry clumps of grass. When they find one, as many as 5,000 individuals cover the clump, each one chewing the grass into thin, half-inch-long strips to be carried back to the nest. The grass is stored underground as the termites built up a food supply to get them through the winter.

But this African termite's habit of harvesting and storing food for winter is one that humans supposedly did not evolve until only recently in their supposed evolutionary history. How can anyone believe that termites could figure out how to harvest and store food before humans did? Even the termite tells us we were created.

Prayer: Dear Lord, I thank You that Your creation is filled with so much intelligence. Take my intelligence into Your service, refine it, and make me better able to witness Your truth and love in Jesus Christ to those around me. Amen.

An Important Extra Bone

Job 23:12-13
"Neither have I gone back from the commandment of his lips; I have esteemed the words of his mouth more that my necessary food. But he is in one mind, and who can turn him? and what his soul desireth, even that he doeth."

How is it that a 50-pound sea otter can rip an abalone – its favorite food – off a rock, while a 150-pound human has a hard time cracking a simple oyster shell? The answer calls into question those evolutionary charts that compare the hand and paw bones of various mammals, including humans.

Similarity of structure doesn't mean that two structures are related. Those tiny hairs in your ear that pick up vibrations so you can hear are nearly identical to the cilia used by some one-celled creatures to propel themselves. But no one has ever suggested that our ears evolved from microorganisms.

And sometimes body structures are not as similar as evolutionists think. Using a CT scanner and an "image computer" originally designed to create special effects for Star Wars movies, scientists have discovered why otters can easily handle shellfish that pose problems for a human three times the otter's size. The sea otter's wrist has an extra bone in just the right place to give the otter an amazing amount of leverage in grasping shellfish. The bone was discovered quite by accident. Scientists weren't looking for it, since their evolutionary beliefs didn't lead them to expect the extra bone.

The sea otter is an example of how God has specially designed each creature. Sure, there are many similarities between creatures, since no good engineer will waste his time trying to reinvent the wheel. But there are also unique differences between creatures that support the Bible's account of creation.

Prayer: Dear Father, not only have You created many unique creatures, You have even made each individual a unique person. Help me to remember that You created me the way I am – minus sin – because You wanted to have a close personal relationship with me. In Jesus' Name. Amen.

Ref: "Why abalones don't find otters cute." *Discover*, Apr. 1988. p. 10.

The Promise of Poison Frogs

Isaiah 38:17
"Behold, for peace I had great bitterness: but thou hast in love to my soul delivered it from the pit of corruption: for thou hast cast all my sins behind thy back."

The tiny poison dart frogs of the American tropics are helping scientists to understand just how nerves and muscles work. Yet the poisons of these frogs are among the most toxic substances known to humans. The poison from one individual, less than an inch long, can kill 50 human beings.

The brightly colored creatures are called poison dart frogs because Indian hunters have tipped their blowgun darts with the poisons for centuries. The 50 to 120 known species of frogs produce over 200 kinds of deadly poisons, which are secreted through their skin. The poisons work by shutting down the electrical charges that make muscles and nerves operate.

Scientists believe they may be able to process the poisons into potent painkillers or heart stimulants. Yet these substances may not have the side effects that current painkillers and heart stimulants produce. In addition, scientists believe the poisons can help them understand disorders like Alzheimer's disease and abnormal heart rhythms.

It may be possible that before sin infected God's perfect creation, the poisons of these deadly frogs were slightly different chemically, making them health-producing medicines. Perhaps human efforts to subdue the earth could restore these substances to their former healthy state. But such efforts can only reverse the earthly, temporary effects of sin. Only Jesus Christ as our Lord and Savior can permanently reverse the eternal effects of sin.

Prayer: Dear Lord, I ask that You would bless man's efforts to better his life here and yet use Your people, including me, to help the world understand that our only true eternal hope is found in Your innocent suffering and death for our sin. Amen.

Ref: Brody, Jane. 1990. "Using toxins from tiny frogs, researchers seek clues to disease." *The New York Times*, Jan. 23. p. C3.

Language Studies Tell Us About Ourselves

Romans 10:17
"So then faith cometh by hearing, and hearing by the word of God."

Some scientists who study language are learning that language is far more complex than they had expected. Some linguists have concluded that language is even too complex to be completely learned. They suggest that children already have some principles of language embedded in their brains when they are born.

For example, how would you turn the statement "The man is here" into a question? Even a child knows to move the word "is" to the beginning of the sentence to make it say "Is the man here?" But how do you turn the statement, "The man who is tall is here" into a question? A child who followed the principles used in the last example should come up with, "Is the man who tall is here?" But that makes no sense, and children do not make this mistake. Even children know instinctively to move, not the first verb in this case but the main verb, to yield, "Is the man who is tall here?"

This exercise seems simply enough to us. And that's the point: The rule that governs this seemingly simple solution is difficult to express either in formal linguistic terms or in a computer program. Yet children instinctively solve this problem without ever being clearly taught any rules.

Human beings are unique. Saying that apes can learn language because they can learn some simple signs is like saying that humans can fly because they can jump. Linguistics is yet another area of knowledge which confirms what the Bible says about man being specially made and very different from the animals.

Prayer: Dear Lord, the Word made flesh for my salvation – I thank You for the gift of language so that I can hear and believe Your Word and, through it, grow closer to You. Let me not take this gift for granted or neglect it. Amen.

A Giant Wave of Water

Genesis 7:19-20
"And the waters prevailed exceedingly upon the earth; and all the high hills, that were under the whole heaven, were covered. Fifteen cubits upward did the waters prevail; and the mountains were covered."

What's the biggest wave you've ever seen on a lake or the ocean? Some scientists are now saying there is evidence that in ancient times a wave over 1,000 feet high swept over parts of the Earth, creating some of its geological features.

While a 1,000-foot wave sounds big – and indeed it is – it's not unbelievable. The largest wave ever recorded in modern times was a 1,700-foot-high wave that struck Lituya Bay in a remote area of Alaska. A landslide caused the wave, which was magnified by the shape of the bay until a wall of water almost one-third of a mile high swept onto the opposing hills, stripping them of forests. Two people drowned, but miraculously fisherman aboard two boats were swept out to sea and lived to tell about the ride of a lifetime.

Some geologists have proposed that a similar wave, caused by an underwater landslide, swept across the Hawaiian Islands in the more distant past. As evidence, they point to blankets of gravel, sometimes 25 feet thick, as well as other debris from the ocean that is scattered across the southern slopes of Oahu, Molokai, Lanai and Maui, as well as the west side of Hawaii. Other geologists believe these are the remains of a sea that was once much higher.

According to the Bible, they may both be right. It's possible that these gravel beds and ocean bottom debris may represent both the violence of the great Flood and the higher sea levels at the end of the Flood that is recorded in Genesis.

Prayer: Dear Father, through Peter You warned us that in the last days there would be those who would forget Your judgment in the great Flood. Help Your people, starting with me, understand the day and age in which we are living and do Your work while it is yet day. In Jesus' Name. Amen.

Ref: "Kowabunga, the surf was up!" *Discover*, Feb. 1985. p. 9.

A Busy Protein

Romans 5:8-9
"But God commendeth his love toward us, in that, while we were yet sinners, Christ died for us. Much more then, being now justified by his blood, we shall be saved from wrath through him."

Blood is made up of many components. Red blood cells carry oxygen to all parts of the body. And even though hemoglobin is based on iron, the healthy body, protected by rust inhibitors, doesn't rust. White cells and other blood components fight infection. It's difficult to decide which of the components of blood are more amazing in their design.

But if awards were given to the various components of blood for the amazing ways in which they work, serum albumin would certainly be in the running. Like one of those gadgets advertised on cable television, serum albumin seems to do more things in the bloodstream than seems possible.

Serum albumin, which is made by the liver, regulates the volume of your blood, helping to keep your blood pressure healthy. It also controls your blood concentration so that all the components are there in the right concentration. It stores molecules and moves them to where they are needed when they are needed. When you take an aspirin or a drug, serum albumin sees to it that the medicine gets to where it is needed. Serum albumin also prevents an enzyme that digests proteins in your digestive tract from digesting proteins in the rest of your body.

Indeed, as the Bible says, much of our physical life happens in our blood. But eternal life is only found in the blood of Jesus Christ, shed on the cross in our place – His death in our place – for the forgiveness of our sins before our Creator.

Prayer: Lord, I thank You for the miracle of life. I especially thank You for the gracious and free gift of eternal life through the forgiveness of sins. When I am tempted to comfort myself with my own good life, remind me of my constant need for cleansing in your blood. Amen.

Ref: Amato, I. 1989. "Serum albumin seen in three dimensions." *Science News*, June 10. p. 359.

Genetic Music

Psalm 65:13
"The pastures are clothed with flocks; the valleys also are covered over with corn; they shout for joy, they also sing."

The entire creation is woven together on an invisible fabric of mathematics. Since mathematics is also the basis of music, the creation is often more filled with music than we think. In fact, Christians ought to think twice before dismissing as figurative those Bible passages about the whole of creation praising its Creator.

Sometimes the music of the creation in praise of God is more clearly seen. As one California geneticist was studying the structure of DNA, it struck him that the pattern of bases out of which DNA is built followed a pattern similar to a melody. So he assigned notes to each of the four bases: cytosine for do, adenine for re and mi, guanine for fa and sol, thymine for la and ti, and cytosine again for do.

Upon playing different gene structures according to this orchestration, the geneticist found that rather than endless repetition and wandering melodies, each gene has its own musical style and interesting melody. One collagen gene sounds like Bach. Genes that code for cell adhesion molecules sound like Debussy. And a mouse gene for RNA sounds very much like Chopin.

As our knowledge of the creation grows, so does the witness to God's power and wisdom. Truly the whole creation, in its every part, sings the praises of its Creator. Let us join that praise with our own voices.

> ***Prayer: Dear Father, I thank and praise You in wonder for the work of Your hands. Let my thanksgiving never cease until the day I see You face to face. Amen.***

Ref: "Play the right bases, and you'll hear Bach." *Discover*, Mar. 1987. p. 10.

Your 20-Watt Brain

Psalm 119:11
"Thy word have I hid in mine heart, that I might not sin against thee."

Picture in your mind the sight and sounds of popcorn popping. As you picture the popping becoming more frantic as the popper fills up, do you begin to smell the popcorn? That's part of the wonder of the brain. The brain cannot only store words and ideas, but sights, sounds and even smells.

The average person's memory is able to retain about 100 billion bits of information – the information found in 500 sets of encyclopedias. But to use computer language, the brain is not only a place where information is stored, it is also an information processor. Yet it only weighs a little less that four pounds and uses about 20 watts of energy. Our most sophisticated modern computers don't even begin to approach such efficiency! Incidentally, that memory of popcorn that we stirred up at the beginning of the program used only one-tenth of the amount of energy in one particle of visible light.

Research has shown that the more you use part of your brain, the larger that part becomes – just like building muscles. And if you don't use part of your brain, it starts shrinking. Few of us have developed our ability to memorize things to any great extent. But to show you what can be done, in May of 1974 a Burmese man recited, from memory, 16,000 pages from a Buddhist religious text!

What are you doing with that marvelous organ, the brain, that your Creator gave to you? How much of His word have you stored in your memory as a treasure which can never be taken away?

Prayer: Dear Heavenly Father, I thank You for the wonderful gift of the brain You have given me. Forgive me for thinking that my brain isn't as wonderful as it truly is, and for comparing it to others. Help me to develop the gifts You have given me for Your glory. In Jesus' Name. Amen.

Peanut Butter and Green Diamonds

Isaiah 40:28
"Hast thou not known? hast thou not heard, that the everlasting God, the LORD, the Creator of the ends of the earth, fainteth not, neither is weary? there is no searching of his understanding."

Perhaps you have recently read one of the many news articles about an amazing new process for making diamonds. The process uses simple heat to make diamonds out of anything that has carbon in it.

Diamonds are made of carbon atoms – that black stuff in soot – that are tightly arranged into the careful rows of a crystal. The result is a clear crystal, harder than anything on Earth. Artificial diamonds have been made using a great deal of heat and pressure. But the new process, using temperatures as low as 250° F, condenses carbon atoms out of vapor in low pressure. The new process is much cheaper and is able to place a diamond coating on various materials. Long-wearing machine parts can be diamond coated, as can ball bearings, making them permanently lubricated. Watch crystals, coated with diamond to make them scratch resistant, are currently being marketed by Seiko.

Now, diamonds can be made cheaply and easily out of almost anything. One researcher even made diamond out of peanut butter, although he reported that the diamonds had a greenish tinge to them because of the nitrogen in peanuts.

We have only begun to scratch the surface, so to speak, in learning about the wonderful properties the Creator built into the material world. We are only beginning to discover the depths of His wisdom. Such discovery is in keeping with His instructions to us to subdue the Earth.

Prayer: Dear Father, in amazement I thank You for the wonderful things You have made, only a little of which we have learned so far. I pray that more Christian young people would go into science careers, remaining faithful to the truth of Your Word and faithful to Your charge to subdue the Earth. In Jesus' Name. Amen.

Ref: Amato, Ivan. 1990. "Diamond Fever." *Science News*, v. 138, Aug. 4. p. 72.

Shrimp Kidnaps Chemist!

Job 35:10-11
"But none saith, Where is God my maker, who giveth songs in the night; Who teacheth us more than the beasts of the earth, and maketh us wiser than the fowls of heaven?"

High drama takes place even among the tiny, almost unnoticeable creatures in even the most remote parts of the world. In the frigid waters of Antarctica there's a tiny plankton that kidnaps even tinier chemists for its protection.

The tiny shrimp-like plankton are not much larger than a grain of rice. They spend their days foraging the cold waters for food and hopefully not running into any plankton-eating fish. The tiny creature has absolutely no defenses of its own. But this creature is not a shrimp without a plan. As it searches about, it may find an even smaller snail-like pteropod. If it finds one, it will fasten the pteropod to its back, not harming it at all.

The pteropod makes a chemical that fish hate. Plankton-eating fish that have tried to eat one of the shrimp-like creatures with its chemist backpack have been seen to shake their heads violently and spit the pair out. But the shrimp's captive backpack slows down his foraging for food. So after several days, the shrimp releases the pteropod unharmed to search for more food and, hopefully soon, another pteropod.

It sounds rather silly to suggest that one day, many years ago, one of these tiny plankton noticed that fish never ate pteropods because of the chemical defense they produce. Who would suggest that once he told his friends, all the shrimp-like plankton started kidnapping pteropods? This strategy was obviously taught to the plankton by the Creator Himself.

Prayer: Dear Lord, there is nothing that You have made that You do not care about. Fill me with the same love and concern for what You have created. Help me to love my fellow human beings as You do. Amen.

Ref: Amato, Ivan. 1990. "Kidnapped plankton shares its defenses." *Science News*, v. 138, Aug. 4. p. 69.

Caterpillar's Diet Choices

Psalm 86:8-10
"Among the gods there is none like unto thee, O Lord; neither are there any works like unto thy works. All nations whom thou hast made shall come and worship before thee, O Lord; and shall glorify thy name. For thou art great, and doest wondrous things: thou art God alone."

The diamondback caterpillar is the number one pest of plants in the cabbage family. But scientists are finding that this caterpillar isn't as guilty of causing low yields in the cabbage family plots as had been thought.

It turns out that, given a choice between a cabbage field that is doing very well and a field next door that's not doing so well, the diamondback caterpillar will prefer to munch on the less healthy field where yields are already decreased. That might seem like poor judgment on the part of the caterpillar. But scientists have only recently learned what the caterpillar seems to have known all along.

The caterpillar's primary predator is a parasitoid wasp. But this wasp looks for its meals in healthy cabbage fields. So while the wasp is hunting caterpillars in healthy fields, the caterpillars are in the field that isn't doing quite so well. This means that the solution to caterpillar problems on cabbage family plants is not more insecticide but healthier plants. Researchers have found that by raising the nitrogen level in a cabbage crop by as little as 3-5%, caterpillars are discouraged and wasps are invited to look for any remaining caterpillars. The single step of fertilization by nitrogen not only controls insects and cuts losses but also improves productivity.

The logical question is, how did the caterpillars figure out how to outsmart the wasps? The obvious answer is that their Creator, who knows all about wasps and caterpillars, built this wisdom into the caterpillars.

> ***Prayer: Dear Father in heaven, help us all to learn more about how Your creation works so that we may benefit from Your wisdom and become more productive and less destructive. I especially ask that You would use Your people to lead the way in this so that You may be glorified even among unbelievers. In Jesus' Name. Amen.***

Ref: Weiss, Rick. 1990. "Wormy cabbage: blame the victim." *Science News*, v. 138, Aug. 11. p. 93

A Frog's Creative Broadcasting Solution

I Corinthians 14:33
"For God is not the author of confusion, but of peace, as in all churches of the saints."

If you live in a large city, especially one that is surrounded by other large cities, you know the radio dial can become very crowded. Sometimes radio stations overlap, although unintentionally. To prevent the radio dial from turning into incomprehensible gibberish, radio stations are assigned a frequency on which to operate.

This orderly arrangement is also found among many animals. Male birds often stake out a territory from which they offer their mating call so that they do not interfere with other males of their species. Frogs, too, have mating calls, but often many species of frogs are found in the same swamp. They avoid interfering with each other by each using a unique frequency for their calls. Otherwise no one's mating call could be clearly heard. This arrangement is evidence enough of the Creator's hand.

But there are two South American frogs, often found in the same location, that use the same frequency. How can they do that? One frog calls only above water while the other frog uses his call only under water. Since the surface of the water is an effective barrier for sound, neither call interferes with the other.

Frogs could have never gotten together and agreed on limiting themselves to unique frequencies. For one thing, frogs don't understand the physics of the problem. What should be clear, especially in our technological age, is that the Creator set up this arrangement for the good of everyone.

Prayer: Dear Lord, You know the value of order in our lives. Sin produces chaos and disorder and makes life difficult. Fill my life with Your order through the instruction of Your Holy Word and replace the chaos caused by foolishness with Your orderly instruction and help. Amen.

Ref: Weiss, Rick. 1990. "Frog finds empty bandwidth, then croaks." *Science News*, v. 138, Aug. 11. p. 93

Why Does It Rain?

Job 36:27-29
"For he maketh small the drops of water: they pour down rain according to the vapour thereof: Which the clouds drop and distil upon man abundantly. Also can any understand the spreading of the clouds, or the noise of his tabernacle?"

Despite the advances of modern science, scientists still don't know why it rains. Remember those science lessons in grade school that supposedly explained why it rained? We learned about temperature, dew point and how moisture in a cloud could condense around a tiny ice crystal to form a drop of rain. But scientists are not really all that sure how raindrops form.

The problem is that the 100 million particles of mist in each cubic meter of cloud are all negatively charged. Since their charges are all the same, mist particles don't attract one another; they repel one another. If those conditions stayed the same, it would never rain.

Some scientists have suggested that radon gas is responsible for changing this state of affairs in the cloud so that rain is formed. As the radon decays, absorbed by the droplets, it adds positive charge to them. Because their charges are different than the surrounding negatively charged droplets, they begin to attract each other and soon the droplet has become large enough to fall as a rain drop. Other scientists, unsure of this theory, are staying with the ice crystal theory.

Modern science often offers its conclusions with a very sure-sounding voice. But the fact is, scientists are unsure of many things, including the simple question of why it rains. That's why we shouldn't get too worried when some scientist claims to have discovered something that he says proves the Bible to be in error.

Prayer: Dear Father in heaven, I am amazed at the mysteries surrounding a simple thing like rain. Give me and all of Your people a firmer faith and understanding of Your truth so that we are less easily influenced by the proud claims of finite men. In Jesus' Name and for His sake. Amen.

Ref: "Rain made easy." *Science 84*, May. p. 7.

Plants Focus Light

Psalm 1:1-3
"Blessed is the man that walketh not in the counsel of the ungodly, nor standeth in the way of sinners, nor sitteth in the seat of the scornful. But his delight is in the law of the LORD, and in his law doth he meditate day and night. And he shall be like a tree planted by the rivers of water, that bringeth forth his fruit in his season; his leaf also shall not wither; and whatsoever he doeth shall prosper."

In many ways, a plant is like a machine which changes the energy of sunlight into food energy that is needed by the rest of the living world. The energy which runs the plant is light.

Now, if you were going to design a better, more efficient plant, how might you do it? Obviously, one way to improve a plant would be to create one which can do its job with less light. But remember, the plant cannot store more energy that it receives. And if the plant was too sensitive to light, a strong dose of sunlight could kill it. The obvious solution is to outfit the plant with tiny lenses which focus even dim light into the chemical centers where photosynthesis take place.

Due to our ignorance, modern science has been able to get away with thinking of plants as though they were simpler, unsophisticated forms of life. Now, scientists have discovered that many plants do indeed have special cells which focus and concentrate the available light into the very centers where it is needed. One plant's shaded leaves were found to intensify available light by 26 times. And if the light gets so strong that it could damage the leaves, the light is focused onto cells which can absorb the extra light.

Plants are much more complex, high-performance living things than we ever suspected. Plant design incorporates designs based on a highly sophisticated knowledge of both chemistry and physics. What a wonderful witness they are to our Creator!

Prayer: Dear Lord, there is nothing in the creation which is poorly made – You have made all things with such insight, care and planning. Help me to better understand that You would likewise be even more involved in my life beginning with the forgiveness of sins. Amen.

Traces of What Man Once Was?

I Corinthians 13:12
"For now we see through a glass, darkly; but then face to face: now I know in part; but then shall I know even as also I am known."

What was man like as originally created – before sin started taking its toll on his mind, body and the world in which he lives? The Bible indicates that Adam had knowledge we could never understand today, when we can only, as St. Paul says, "…see in a mirror dimly…"

Are there hints around us of the original capabilities of the human brain? Keep in mind that each of us has different deficiencies and different strengths. The so-called idiot savant has low I.Q., yet is capable of incredible mental feats. This syndrome is caused when damage to the left hemisphere of the brain causes the right hemisphere to "overdevelop."

"K" at the age of 28 had a mental age of 11 and vocabulary of only 58 words. But he could remember the population of every US city of more than 5,000 people, the distances of every city between New York and Chicago, every county seat in the nation and the names, locations and number of rooms in over 2,000 hotels. In the 1700s, Jedidiah Buxton, with a mental age of 10, could calculate any number in his head. When asked how many 1/8ths of an inch exists in a cube 23,145,798 yards by 5,642,732 yards by 54,965, he answered with the correct 28 digit number. Blind and mentally retarded, Leslie Lemke is a musical genius. He can play any music after only hearing it once. He once played every note of a 45-minute piece perfectly after hearing it the first time!

Do we see here a trace of abilities which God meant for all?

Prayer: Dear Father, I confess that I have not made full use of the abilities You have given me, and I cannot even understand the abilities I might have had if we had not sinned. I pray, give me wholeness in the Resurrection. In Jesus' Name. Amen.

Amoeba Society

Romans 8:19
"For the earnest expectation of the creature waiteth for the manifestation of the sons of God. For the creation was made subject to vanity, not willingly but by reason of him who hath subjected the same in hope, Because the creature itself also shall be delivered from the bondage of corruption into the glorious liberty of the children of God."

It's not exactly a plant, but it's not completely an animal either. It crawls like an animal and then grows fruits and sows seed like plants on colorful stalks. Sometimes it acts like a single-celled creature with a refined social order. You can even see each single living individual cell under the microscope. And sometimes all those single living cells fuse into one giant organism.

Scientists aren't sure whether the strange and fascinating form of life which goes by the inglorious name cellular slime mold is a plant or actually an animal. This often beautiful form of life is very different from true slime molds which are rather drab and disgusting to many people.

In the normally moist forest floor, single amoebas – invisible to the naked eye – live in the moisture and food provided by the decaying vegetation. But if the floor of the forest begins to dry out or food is scarce, the amoebas congregate. As they come together, they combine into a tiny slug, about a millimeter long and large enough to see. This apparently individual creature crawls to more favorable conditions and stands up. Then single amoebas are again evident, crawling up, growing off the colorful stalk of other amoebas, until they finally form a mass of spores at the top. These spores "hatch" into individual amoebas when conditions are again right for them.

There is no simple form of life. Even cellar slime mold shows the Creator's touch in its wonderful complexity and amazing way of life.

Prayer: Lord, You have made nothing that is poorly made or worthless. You love everything in Your creation. I thank You that You loved me so much that You gave Your Life in my place for my sin, so that I could be restored to God through the forgiveness of sins. Help me to live as Your redeemed creation. Amen.

One Foot and a Sail

Psalm 104:24-25
"O LORD, how manifold are thy works! in wisdom hast thou made them all: the earth is full of thy riches. So is this great and wide sea, wherein are things creeping innumerable, both small and great beasts."

According to unsubstantiated reports, it was about 50 years ago that Chinese immigrants to the United States brought some fresh water clams with them. The clams were eaten for good luck on Chinese New Year. What is substantiated is that today the clam is found in at least 35 states. That's an amazing amount of territory to cover in only 50 years for creatures that only have one foot.

The spread of the Asian clam costs industry over a billion dollars per year in clean-up costs. Lacking enemies, it reproduces rapidly and clogs the water intake line which industry uses to cool its equipment. Scientists began to wonder how the creature could travel so rapidly, even going from lake to lake and stream to stream.

What scientists learned was that when the Asian clam encounters a current, it sets sail. Lifting off with its one foot, it releases a transparent film that catches the current like a sail. It sails off, using its foot like a rudder. This explains not only why it spread so rapidly down rivers, but how the clam could spread from one body of water to another during floods.

The Asian clam's ability to sail is a clever design. So, too, is the balance of nature which prevents creatures from over-populating. Unfortunately, the Asian clam in the United States has few enemies to keep it under control. If evolution were true, such lack of balance would be the rule in nature. The balance that exists, keeping each creature fed but not over-populated in their natural settings, is but another witness to the fact of the Creator.

***Prayer:** Dear Heavenly Father, the balance You have designed between all living things, so that they can make their living and yet do not run out of control, shows Your continuing love toward the fallen creation. Help me to be a clear and specific witness to Your love for us in Jesus Christ. Amen.*

A Not So Rock-Solid Moon

Philippians 2:5-7
"Let this mind be in you, which was also in Christ Jesus: Who, being in the form of God, thought it not robbery to be equal with God: But made himself of no reputation, and took upon him the form of a servant, and was made in the likeness of men:"

How old is the moon? Evolutionary scientists talk about the craters on the moon being sixty-five or a hundred million years old or even billions of years old. But since we have learned what kind of rock the moon is made of, we now know that these ages are impossibly old.

Craters are made when a large rock from space crashes into the moon at high speed. If the moon were made of water, a rock would only make a big splash. That splash would be a crater for only a few seconds before it was once again filled with water. If the moon were made of honey, the crater would last a little longer, but the honey would soon flow smooth again and the crater would be gone.

American astronauts returned to the Earth with many samples of the moon rock for scientific study. The basic rock on the moon is basalt. Studying this rock, scientists have learned that basalt, too, flows. The rate of flow is slower than that of honey or even molasses in January, but it can be measured. According to their measurements, none of the craters on the moon can be more than a few thousand years old. The natural tendency of the basalt to flow would have erased any craters which are older than that.

Science has shown that the moon craters which evolutionary scientists thought were millions and billions of years old are really only thousands of years old at most. This is one more piece of evidence that the Bible's explanation of a young creation is not only true but can be scientifically supported.

Prayer: Dear Lord, You are the Maker of all things in the universe. When I think of the size of the universe and the tremendous energies You created in space, I am even more deeply moved that You became one of us so that I could have the forgiveness of sins. Truly, You are worthy of all worship! Amen.

Whales: Armed and Dangerous

Psalm 148:7
"Praise the LORD from the earth, ye dragons, and all deeps:"

The narwhal is a small whale, reaching perhaps 10 feet in length, which has an eight-foot spiral tusk coming out of its head. While a whale with a tusk nearly as long as its body might be a dramatic sight, it must be difficult for the narwhal to hunt down its food. The strap-toothed whale would seem to have an even more difficult time catching it prey. You see, this whale has two long teeth growing out of its lower jaw which curve over its upper jaw, preventing it from opening it mouth very wide.

Now scientists believe that they may have discovered how these apparently ill-equipped creatures, along with the other toothed sea mammals, make their living. They suspect that toothed whales and dolphins can create explosively loud, focused sound beams underwater and literally stun the fish they eat.

Bottle-nosed dolphins have been seen to emit sounds louder than 228 decibels, nearly the same as a blasting cap under water. It seems they quite literally catch fish with dynamite! Sperm whales can outdo a blasting cap, generating 265 decibels.

The wonders to be discovered in the living world beautifully combine and harmonize even the most subtle principles of chemistry, physics and biology in most creative and imaginative ways. What science is learning leaves us with no other conclusion than the Bible's own statement that we have a wonderful, powerful and glorious Creator.

Prayer: Dear Father, one of the greatest mysteries is how men can say that there is no Creator. I ask that You would move more Christian young people to enter the scientific professions and never waver in their witness to You. In Jesus' Name. Amen.

Economic Life Among Bumblebees

II Thessalonians 3:10
"For even when we were with you, this we commanded you, that if any would not work, neither should he eat."

What a beautiful scene. A field filled with a riot of colors, with bumblebees lazily going from flower to flower. To us this is a relaxing scene of idyllic peace. The fact is, the pace for the bee is as hectic as for any human commuting in rush hour traffic. This is largely because, like humans, bumblebees are warm-blooded.

Many books about bee societies extol the virtues of communism. These books almost make it appear as if bees sit around deciding who should work which flowers, all for the good of the hive. But it is precisely because this method is less productive that warm-blooded bumblebees, who need more energy, must use a system that requires personal initiative to increase productivity.

Researchers have found that bumblebees gather nectar and return it to the hive completely on their own initiative. Young bumblebees gather nectar from many different kinds of flowers until they find which flowers produce the most nectar. When the most rewarding flowers dry up, they switch to the second most rewarding. So individual motivation leads each bee to make the most of his own efforts. As a result, the entire colony benefits.

One researcher commented that this method works so well that it almost looks like the entire hive is guided by an "invisible hand." This is more than simply appearance, for God does care for all His creatures, including you and me. He has shown His greatest love for us in sending His Son, Jesus Christ, so that we could have the forgiveness of sins through Him.

Prayer: Dear Lord, I thank You that You left the Father's presence in heaven to take our humble form upon Yourself. I believe Your Word which tells me that You have carried my sin so that, through forgiveness, I could stand cleansed before my Creator. Amen.

Ref: Jordan, William. 1984. "The bee complex." *Science 84*, May. p. 58.

Facts that Weren't

I Timothy 6:20-21
"O Timothy, keep that which is committed to thy trust, avoiding profane and vain babblings, and oppositions of science falsely so called: which some professing have erred concerning the faith. Grace be with thee. Amen."

What is the Christian to think when told so often that the "facts of science" prove evolution – and disprove the Bible's story of man's history?

Not all "facts" are created equal. For example, the fact that Jesus Christ was a real person is not open to interpretation. He is mentioned in the Bible and by other writings of the time – that's a fact. Evolutionists use a different meaning for the word "fact."

A fossil, the "Texexpan Man," is an example of an evolutionary "fact." Texexpan Man is a fossil which, school children were taught, lived over 10,000 years ago. Evolutionists themselves recently proved that "Tepexpan Man" is not only less than 2,000 years old but is actually a woman! An object found in the United States was said to be a fossilized egg, possibly 16 million years old. Now it has been learned that the object is a three- to-five-year-old stomach stone from a modern mammal – possibly a cow. A giant fossil found in South Australia early in 1990 was declared to be Australia's largest dinosaur bone. Further study has now shown the object to be a fossilized tree trunk. All "facts" are not created equal.

If you went to school more than 20 years ago, you need to know that most of the "facts" that you were taught in support of evolution have since been found to be either in error or open to serious question. Evolutionary "facts" are only interpretations of the world. So the Christian can follow the Scripture with confidence, knowing that the Bible is unchallenged by the "facts" of evolution.

Prayer: Dear Lord Jesus, as the Word made Flesh, You are the Way, the Truth, and the Life. Give Christians a boldness of faith so that we are not intimidated by the claims against Your truth which are made by the unbelieving world. Amen.

Horned Animals in Australia?

I Peter 3:18-20
"For Christ also hath once suffered for sins, the just for the unjust, that he might bring us to God, being put to death in the flesh, but quickened by the Spirit: By which also he went and preached unto the spirits in prison; Which sometime were disobedient, when once the longsuffering of God waited in the days of Noah, while the ark was a preparing, wherein few, that is, eight souls were saved by water."

Australia is known for its unusual animals. But not only are there animals which are found only in Australia, many animals commonly found elsewhere in the world are not found in Australia. For example, no animals with horns are found in Australia.

The fact that no animals with horns are known in Australia makes it difficult to explain aboriginal rock art from Mount Manning, north of Sydney. This art, drawn before Europeans settled in Australia, clearly shows an animal with horns!

The obvious explanation is that the first aboriginal people to move to Australia brought with them memories of horned cattle or deer from Asia. But most evolutionary scientists reject this explanation because, according to evolutionary dating, the memory would have had to have been passed down through 2,000 generations. Evolutionists correctly conclude that this is hard to believe.

But if the Earth is relatively young, as the Bible says, and the aboriginal population settled in Australia after the great Flood, much less time is involved. Instead of 2,000 generations between aboriginal and European settlement, there would be less than 150 generations! This is a reasonable number of generations to preserve memories of horned creatures. What's more, it supports the idea that all human beings descended from the original eight Flood survivors and had a knowledge of creatures unknown to the part of the world where these descendants live today.

Prayer: Dear Father in heaven, I give you thanks when I see how the facts of human history from outside the Bible so clearly support the history of man that is related in the Bible. Help me understand more of these facts so that I make a more effective witness to Your truth. In Jesus' Name. Amen.

Yoho Surprise!

Exodus 20:11
"For in six days the LORD made heaven and earth, the sea, and all that in them is, and rested the seventh day: wherefore the LORD blessed the Sabbath day, and hallowed it."

Evolution says that life today arose from fewer, simpler forms of life. Evolutionists tell us that life is an ever ascending ladder of increasing complexity as new forms of life develop. For proof, they claim that the fossil record shows this kind of development of life. But is this really what the fossil record shows?

It would be easy for me, as a creationist, to tell you that the fossil record shows no increase in biological complexity. But a listener recently sent me a brochure from the Yoho National Park in Canada. The brochure talks about the richness and complexity of the fossilized life forms in the Burgess Shale of the park. Published by the Environment Canada Parks Service, the brochure clearly says that the fossils in the park contradict standard evolution.

Let me share a few quotes with you. "Life is simpler today than it was long ago." "Long ago things used to be much more complicated. In fact, this is the central truth underlying one of the most revolutionary discoveries in the history of paleontology." "The animals of the Burgess Shale were more diverse and no less highly-specialized than today's living creatures."

The Bible gives us the same picture as the fossil record. Living things have always been complex. But early in the Earth's history there were many more kinds of living things than there are today. And no fossil anywhere shows one kind of creature changing into another. The Bible has correctly predicted what is now found in the fossil record.

Prayer: Dear Lord, kings and armies have tried to stand against Your Word of life to us. They are gone and Your Word remains. Let me never imagine that any man could successfully challenge Your Word, but let me treasure the Word by using it. Amen.

The Warm-Blooded Bumblebee

Genesis 1:27
"So God created man in his own image, in the image of God he created him; male and female created he them."

What is warm-blooded, sits on its eggs like a hen, but flies much better than a hen, and has six legs? It's the bumblebee.

The amazing fact that bumblebees are warm-blooded was discovered after a scientist saw a bumblebee gathering nectar one frosty morning. As it turned out, even with air temperatures near freezing, a bumblebee can maintain a body temperature of 95° F. In order for the bee's tiny body to stay warm, it has a power output that is 20 to 30 times that of a world-class marathon runner!

Bumblebees are made up of three body sections: the head, the thorax, and the abdomen. They heat their blood in the thorax, where the flight muscles are located and where the heat is most needed. This heat is regulated by the bee's circulatory system, which flows through its narrow "waist." Blood passing through this "waist" from the cooler abdomen can be heated by the sun if necessary. It then passes through a heat exchanger in which it is further heated by the warmer blood flowing away from the muscles. On a hot day, the bee does the neatest trick of all. It actually changes its circulatory system, so that fresh blood coming to the wing muscles is not heated at all by the heat exchange system. The queen bumblebee also uses this ability to control heat to incubate her eggs at 80 degrees, like a mother hen.

There is no simple form of life. What sets humans apart from other life has nothing to do with warm blood, their brains, or evolution. Humans, created in God's image, find redemption only in Jesus Christ.

Prayer: Dear Father in heaven; I thank You that You do not evaluate me for what I am – a sinner. Since I have forgiveness of sins through Your Son, Jesus Christ, I trust Your promise that when You look at me, You see His sinless life. In His Name. Amen.

Ref: Jordan, William. 1984. "The bee complex." *Science 84*, May. p. 58.

Phony Fungus Flowers

Luke 12:25-26
"And which of you with taking thought can add to his stature one cubit? If ye then be not able to do that thing which is least, why take ye thought for the rest?"

As bees go about their business of gathering nectar, they also pollinate many species of flowers, making possible the next generation. But even though it doesn't have flowers, pollen, or nectar, the fungus which causes the mummy-berry disease still fools bees into making its next generation possible.

As soon as the fungus infects the leaves of a blueberry plant, the leaves begin to wilt. But they also do some other amazing things. The infected leaves begin to ooze a nectar-like substance, send out a scent, and reflect light in the ultraviolet range. To a bee, these are the three unmistakable signs of a flower.

Bees will land on the leaves, taking sugar from the oozing liquid and getting some of the fungus spores on themselves. Sooner or later, they will land on a healthy blueberry flower, leaving some of the spores to infect the flower. Blueberry flowers which have been infected with the spores produce white seedless and tasteless mummy-berries which are full of fungus. The fungus lies dormant inside the berries until spring when they infect the new blueberry leaves, starting the cycle all over again.

If the mummy-berry fungus invented this amazing plan to reproduce itself, it would have to have a greater knowledge of biochemistry than modern science and a good knowledge of both the physics of light and bee behavior. If good for nothing else, the mummy-berry fungus teaches us that the creation could not have made itself. The One and Only Creator God made everything!

Prayer: Dear Lord, if You have given such great gifts even to a fungus, how much more do You offer me. You have told us in Your Word, "You don't have because you do not ask." I ask first of all for understanding that I may know what You would have me ask, then for a bold faith so that I will ask. Amen.

Thomas Jefferson Speaks Out

Psalm 139:15
"My substance was not hid from thee, when I was made in secret, and curiously wrought in the lowest parts of the earth."

It may surprise you to learn that evolution goes back to a time long before Charles Darwin. Even two thousand years ago, those who wanted to escape the inevitable conclusion that we are responsible to our Creator were trying to explain all things by appealing to natural laws.

So, though he lived before Charles Darwin, Thomas Jefferson knew of evolutionary ideas because men like Immanuel Kant, James Hutton and Thomas Malthus were promoting them. Of course, with three doctorates, well-versed in the sciences, and at home in six languages, Jefferson was as well-educated as anyone. And he pointedly rejected evolutionary ideas.

Jefferson wrote against evolution and in defense of belief in the Creator. "I hold (without appeal to revelation) that when we take a view of the universe in its parts, general or particular, it is impossible for the human mind not to perceive and feel a conviction of design, consummated skill, and indefinite power, in every atom of its composition.... It is impossible...for the human mind not to believe that there is, in all this design, case, and effect, up to an ultimate cause, a Fabricator of all things.... Surely this unanimous sentiment renders this more probable than that of the few other hypotheses..."

We add that the creation testifies of the Creator of all, but only in the Bible do we learn of His deep love for us in Jesus Christ. Let the creation lead you to His revealed Word and love for you in Christ.

Prayer: Dear Father, when I am moved by a beautiful sunset or the sweet song of a bird to glorify You, let me be reminded to return to Your Word that I may learn even more about the forgiveness of my sins in Jesus Christ. Amen.

The Bird that Lies and Steals

I John 1:9
"If we confess our sins, he is faithful and just to forgive us our sins, and to cleanse us from all unrighteousness."

Many birds live within a society that is much more complex than any ant or bee society. In fact, it's quite amazing how closely some bird societies resemble human societies.

In the Amazon rain forests of Peru, twenty or more mixed species of birds will flock together in the lower levels of the rain forest canopy. Each flock has a leader and an organizer – the bluish-slate ant-shrike. Each morning the ant-shrike will assemble its flock with loud calls.

This strange arrangement has a very good purpose. The ant-shrike watches over the flock like a mother watches her children. When a bird-eating hawk appears overhead, the ant-shrike shrieks a warning so that members of the flock can take cover. But as often as half the time, the ant-shrike is sounding the alarm to distract members of the flock from the insects they have uncovered. The ant-shrike gets up to 85 percent of its food by sounding a false hawk alarm. The rest of the members of the flock put up with this lying and thieving because of the ant-shrike's value as a hawk warning.

Lying and stealing are, of course, wrong for human beings. And our Creator will hold us each personally responsible for our thoughts and actions. But His Son, Jesus Christ, took our sins upon Himself so that all who believe in Him as their Lord and Savior will stand before Him cleansed of all their sins. Our Creator has promised this!

__Prayer: I thank You, Lord, that You have taken my sin upon Yourself. Help my trust to always rest in what You have done for me and not in what I think I can do to better myself before the Heavenly Father. Amen.__

Chimps Discover New Antibiotic!

Romans 8:20-21
"For the creature was made subject to vanity, not willingly, but by reason of him who hath subjected the same in hope, Because the creature itself also shall be delivered from the bondage of corruption into the glorious liberty of the children of God."

When penicillin was first used as an antibiotic, it was considered a wonder drug. Infections which had been fatal for thousands of years were suddenly curable. Without antibiotics, I would not be here today and many of you would not be reading this.

Man has long known that some natural substances have a natural ability to fight infections. But recently, modern medicine had to step aside and acknowledge that not all medical breakthroughs are made by doctors – or even human beings.

For years it has been known that the roots and leaves of the Aspilia shrub are used as a traditional African treatment for wounds and stomach aches. Researchers also observed chimpanzees swallowing the leaves of the shrub. This led them to investigate more about the leaves. They learned that Aspilia leaves and roots contain a chemical which kills some infectious bacteria as well as fungi and worms. Researchers hope to learn whether this chemical is safe and effective for humans when taken orally.

However, the mystery remains. How did the chimps discover the antibiotic nature of Aspilia? We need to look to our Creator Who appears to have given the chimps this knowledge. While disease and death are not part of His original design for the creation, He was loving enough to provide medicinal help for these consequences of human sin. But the permanent cure for sin and death began with the birth of His Son, Jesus Christ.

Prayer: Dear Father, disease and death were not Your plan for the creation, but the result of man's rebellion against You. Let the wide scope of Your mercy in providing us with some earthly remedies for the results of sin fill me with even greater appreciation for Your boundless mercy and love to us in Christ. In His Name. Amen.

The Birth that Ended Death

I Corinthians 15:22
"For as in Adam all die, even so in Christ all shall be made alive."

People ask, "Why is there death and suffering in the world? Is this the way God wants it?" The Bible's clear answer is an emphatic "No!" The world God made had no death, suffering or disease.

Since all our experiences are in the world as it is now, it is difficult for us to even understand what God intended for us. When God created the universe and mankind, man knew God with a perfect knowledge and was happy with that knowledge. God and man communicated with each other more simply and easily than we, in a fallen world, can communicate with each other today. This is what God wanted for man.

But God did not make robots. God gave us the ability to withdraw our love from Him, for only love which is given when there is a choice not to love is meaningful. However, God warned man that to withdraw his love from God meant death. We cannot blame our first parents for the mess we are in today. Each of us has withdrawn our love from God by sinning too. The resulting suffering in this world is only a warning of the eternal consequences of our sin.

The birth of the Son of God in the Person of Jesus Christ was the birth that ended death. In taking our created form upon Himself, He lived a perfect life in each of our places. Then, although He deserved no punishment for sin, He willingly suffered that punishment in our place. His Resurrection from the dead was God's signal that He had accepted Christ's sacrifice on our behalf. In Christ, death has died, and life has been re-established for us!

Prayer: Dear Lord, I confess that I will never understand the depth of Your love for me while I am in this world. But I believe that You have indeed rescued me from sin, death and the devil. Help my unbelief. Amen.

A Computer with Fins

Job 21:22
"Shall any teach God knowledge, since he judgeth those that are high?"

The glass knife fish, popular in home aquariums, has been the subject of study by two University of California researchers. They have concluded that this unusual looking fish has such a complex nervous system that it is basically a computer with fins.

As the six- to seven-inch native of South America swims along, it generates 250 to 700 weak electrical impulses a second, which it uses like radar to navigate. The returning signals are received by specialized cells all over the fish's body.

The fish's brain then processes these signal in much the same way as our brains process the direction from which a sound comes. We sense the direction that a sound comes from because the distance between our ears means that the incoming sound can strike one ear a mere 15 millionths of a second earlier than the other. But electrical impulses travel much faster than sound. So the glass knife fish can actually detect a difference of 400 billionths of a second! The fish is sending thousands of these reception signals to its brain for *each* of the up to 700 signals it transmits every second. The fish's secret to processing this huge amount of information is called parallel processing, a design which has only recently been discovered and applied to computers to create the fastest supercomputers yet.

The glass knife fish has too unlikely a design for evolution to have made it. On the positive side, the sophisticated information processing technology built into this fish can only be attributed to the Creator present to us in the Bible.

Prayer: Dear Father in heaven, man's greatest knowledge and wisdom and his most advanced discoveries are so small in comparison with what You have made that they serve to give us a greater appreciation of what You have done in creation. Remove human pride from me and teach me complete dependence on Your Son, my Lord and Savior, Jesus Christ. Amen.

He Breathes Through His Legs

Isaiah 29:14
"Therefore, behold, I will proceed to do a marvellous work among this people, even a marvellous work and a wonder: for the wisdom of their wise men shall perish, and the understanding of their prudent men shall be hid."

Living things generally have similar parts. This is because those parts which are similarly constructed in different living creature usually do similar things. The ape's arm is similar to man's because they have similar jobs to do.

Evolutionists have used this fact to argue in favor of evolutionary relationships between living things. Logically, of course, simply because two different things have similarities doesn't mean that one led to the other. Even though there are hundreds of automobile manufacturers around the world, cars are always identifiable as cars because they are designed for the same purpose.

God seems to have taught us the lesson that similarities between living things don't mean they are related by filling His creation with novelties. Almost all animals get their oxygen through lungs or gills. These organs are nearly always located on the main part of the body. This rule is so universal that the breathing organs of sand-bubbler crabs have been misidentified by scientists for over a century. Scientists never checked their identification of the disks on the crab's legs as hearing organs until a few years ago. When they did, they found that the sand-bubbler crab actually breathes through its legs! Its hearing organs are located elsewhere!

God's creation of novelties like this in living things should do more than lead us to praise Him in wonder over His creativity. They should also help us challenge ideas and theories which try to deny that He is our Creator.

Prayer: Dear Father, forgive me for the times that I have been intimidated into silence in the face of ideas which deny You. Grant me Your Holy Spirit and wisdom so that I may boldly tell others the truth of Who You are and what You have done for us sinners through Jesus Christ, Your Son. Amen.

Legs Knocked Out from Under Snake Evolution

Genesis 3:14
"And the LORD God said unto the serpent, Because thou hast done this, thou art cursed above all cattle, and above every beast of the field; upon thy belly shalt thou go, and dust shalt thou eat all the days of thy life."

It's rather interesting that evolutionists believe that snakes once had legs and lost them. While strictly speaking, according to Genesis, it was a "serpent" and not necessarily a snake which allowed itself to be Satan's instrument for the temptation of Eve, tradition has always identified the "serpent" with the snake.

Obviously, evolutionists don't accept the story of the first temptation. As far as they are concerned, the snake evolved from some reptile which originally had legs. But evolutionists have always tried to find some benefit to not having legs satisfy evolutionary theory. In 1973, a research study suggested that garter snakes use 30 percent less energy for locomotion than they would if they had legs. That study was preliminary and never published. But that didn't stop evolutionists from saying that they had found the reason that snakes don't have legs.

Now, a much more exhaustive study done at the University of California at Irvine, has shown that this evolutionary explanation is false. Outfitting black racer snakes with oxygen masks and using modern equipment including a snake-sized treadmill, researchers have shown that snakes use as much energy as a legged creature of the same weight to get around. The supposed evolutionary advantage of not having legs has disappeared under the bright light of scientific investigation.

And so yet another so-called scientific claim that the Bible has been proved wrong fails in the light of careful science.

Prayer: Dear Father, modern man seeks to avoid this responsibility to You by trying to explain the world as if You are not its Creator. Correct and call me when I try to deny my sin, for I know that You are love, and that Your love has been shown to me in the forgiveness of sins though Jesus Christ. In His Name. Amen.

Colorful Life in the Goldfish Bowl

Psalm 2:1
"Why do the heathen rage, and the people imagine a vain thing?"

Did you ever notice how your photographs sometimes turn out showing different colors than you remember seeing when you took the picture? Florescent lights might make faces look a little green. A tungsten lamp will make everything in a photograph look pinkish. Even though you did not see those colors when you took the picture, they were there and could be measured by the right equipment.

The reason that the colors we see are often different is because or brains have the amazing ability to adjust and correct the colors we see so that they are more realistic. This ability, known as color constancy, was discovered more than 25 years ago by Edwin Land, founder of the Polaroid Corporation.

More recently, researchers using the same kind of test that Land uses, only with goldfish, discovered that goldfish not only see color but also that their brains adjust for color constancy. Scientists had always thought that goldfish were too primitive and had brains that were too small to even see color, much less provide color constancy.

There is no such thing as a simple living thing. Nor can living things be arranged on a scale from simple to complex, the way many scientists would like. Color vision and color constancy were not produced by trial-and-error genetic accidents; they are abilities given by the Creator. He can give these abilities to any creature He wants – and has done so, that the entire creation might glorify Him!

Prayer: Dear Lord, the wise of this world stand against You and Your revealed truth in their pride, yet they stand in awe of what has been created. Grant that Your people would not be intimidated by the proud, self-proclaimed knowledge of the world, and that your people would more clearly see what can be known of You through the creation. Amen.

A Whale of a Song

Psalm 98:4
"Make a joyful noise unto the LORD, all the earth: make a loud noise, and rejoice, and sing praise."

Some have suggested the existence of a principle of universal mathematics among all living things. There does appear to be some musical principles to illustrate this idea; it is called mathematical Platonism.

Although they might be expected to produce random noise, humpback whales produce rhythms in their songs that are like human music. The phrasing of these songs lasts a few seconds, just as they do in human music, even though they could produce much longer phrases. And like human music, the whales' songs repeat themes. The songs themselves typically run between the length of a human ballad and a symphony. They sing a theme, part of which they embellish, and then they go back to a modified version of the theme. While humpbacks are capable of singing over seven octaves, they actually sing notes that are similar in interval, or tone change, to human music. The whales are also capable of percussion in their songs, and they use it in about the same ratio as found in Western symphonies. Finally, like human music, the songs of humpbacks have repeating sections that create rhymes.

Whether these similarities are or are not evidence of mathematical Platonism, we don't know. But there is no question that the songs of whales glorify their Creator and ours. May we praise our Lord for salvation both here and in eternity.

Prayer: Lord, I praise you for the gift of music. Help me praise You for Your salvation, both here and in eternity. Amen.

Ref: *Science*, Vol. 291, 5/1/01, "The Music of Nature and the Nature of Music."

Stealthy Ants

James 5:20
"Let him know, that he which converteth the sinner from the error of his way shall save a soul from death, and shall hide a multitude of sins."

The *Hirtella* tree of French Guiana is very ant-friendly. A mature tree has hollow pouches at the base of their leaves that make comfortable ant apartments. In addition, it provides a sweet nectar for the ants.

Researchers studying the tree noticed that it typically had patches of gray mold growing on it. Looking a little closer, they discovered ants hiding in a leaf pouch underneath the mold. The ants had their mouths open and were looking out through holes in the mold. Scientists suspected that this was some sort of trap, so they placed various insects on the mold. Sure enough, the ants reached through the mold to hold and sting the insects. The ants' trick even worked with grasshoppers, which are much larger than the ants. Further research has shown that *Hirtella* trees without ants don't have the patches of mold. The ants find the mold, bring it to the trees and cultivate and trim it to make their traps.

The relationship between the *Hirtella* tree and the ants is amazing enough. However, the intelligence behind the ants' stealthy activity is even more amazing and could only have come from an intelligent Creator. That Creator Who taught the ants to do this also sent His Son, Jesus Christ, to teach us how to live and how to be reconciled back to our Creator.

Prayer: Father, thank You for forgiving my sins through the saving work of Your Son and my Lord, Jesus Christ. Amen.

Ref: *Science News*, 4/23/05, p. 260, S. Milius, "Ambush Ants."

The Strangest Genetic Relationship

Daniel 4:3
"How great are his signs! and how mighty are his wonders! His kingdom is an everlasting kingdom, and his dominion is from generation to generation."

A very strange type of virus, that some scientists don't even think is a virus, depends upon a parisitoid wasp for its existence. Without the wasp, the virus cannot reproduce, nor can the wasp reproduce without the virus.

The wasp is referred to as a parisitoid wasp because it kills its host, whereas a true parasite merely controls its host. The drama begins almost unnoticed by a caterpillar as the small wasp stings it. Ten days later the caterpillar stops moving but stays attached to whatever plant it's on. Then the eggs that the wasp had injected with its sting begin to hatch and the hatchlings begin eating their way out of the caterpillar. The question is, why didn't the caterpillar's immune system destroy the eggs? The secret is that when laying the eggs, the wasp also injected the unusual virus that shuts down the caterpillar's immune system. The only place the virus can reproduce is inside the wasp, where it does no harm. In fact, the relationship between the virus and the wasp is such that the DNA of the virus has actually become part of the genetic structure of the wasp.

The relationship between the wasp and the virus is so close that it is obviously created. The evidence for this is especially clear since the virus is genetically dependent on the wasp.

Prayer: Lord, Your wonders are beyond counting, especially the wonder of Your plan of salvation through Your Son. Amen.

Ref: *Science News*, 2/26/05, pp. 136-137, David Shiga, "Poisonous Partnership."

Better Flower Power

1 John 1:8
"If we say that we have no sin, we deceive ourselves, and the truth is not in us."

Have you heard about orchids that can fool bees and other pollinators so that they can reproduce? Actually, several hundred orchids have been discovered that fool pollinators with fake sexual signals. However, the champion deceiver has only recently been discovered, and it works its deception not only in a better way, but in a different way from all the others.

The orchid species, known by its Latin name as *O. speculum*, looks something like the female of the wasp species that pollinates it. However, its secret weapon is its scent. While most orchids that engage in such deception mix over a dozen chemicals to create the scent of the female of the species that pollinates it, this new orchid specializes. It uses only ten chemicals, all of them rare, to create the scent which the wasps respond to. Only one of these chemicals has been found before in honeybees. And while the other orchids might attract only inexperienced pollinators, this new orchid will attract even experienced wasps. It does this by doing a better job of creating the wasp attractant than a female wasp can.

Obviously, orchids know nothing about chemistry, and could not study the wasps before it needed a pollinator. This arrangement is clearly the work of a wise Creator, Who created both the wasps and the orchid.

***Prayer:** Father, help me not to be deceived by sin, but always place my trust in Jesus Christ for forgiveness. Amen.*

Ref: *Science News*, 2/1/03, pp. 67-68, S. Milius, "Better Than Real."

The Jellyfish with 24 Eyes

Psalm 139:24
"Thine eyes did see my substance, yet being unperfect; and in thy book all my members were written, which in continuance were fashioned, when as yet there was none of them."

Many writers have pointed out that the claim that vision evolved gradually, over time, is difficult to believe. After all, there are many parts to the eye – there is the lens, the optic nerve, and there must be a portion of the brain that interprets what is seen. If any of these parts doesn't work, there is no vision and no survival advantage.

However, evolutionists not only claim that vision evolved, but that it evolved several times. Because their vision systems are so different, evolutionists believe that vision evolved separately in cephalopods, vertebrates and the box jellyfish. Box jellyfish are better swimmers than most jellyfish and some even have a courting ritual. A member of the family of box jellyfish has 24 eyes, six each on four different stalks. New research at Lund University in Sweden has shown that these eyes appear to be specialized. Eight of the eyes, two on each stalk, have lenses. Other eyes have irises that respond to light. Scientists believe that, despite all of these eyes, the box jellyfish sees only a blurry picture of its surroundings. However, it is able to see it in a wide-screen version.

Vision is a great blessing from God, and a witness that all things are His handiwork. Some of His wonders have been formed so that we make no mistake about their origin.

Prayer: I thank You, Lord, that You have created such wonders. Help my life to glorify You, too. Amen.

Ref: *Science News*, 5/14/05, pp. 307-308, S. Milius, "Built for Blurs."

Letting God Create Your Day, Volume 1 Index

Page	Title
55	100-Foot Ferns
47	A 2,000-Year-Old Computer
52	A 2,000-Year-Old Electric Battery
67	A 4,000-Year-Old Computer Language
227	A Busy Protein
251	A Computer with Fins
82	A Confused Flower?
156	A Desert Traveler's Friend
58	A Diet of Oil
39	A Disastrous Evolutionary Explanation
233	A Frog's Creative Broadcasting Solution
226	A Giant Wave of Water
61	A Journey Through Inner Space
63	A Monkey, A Bird and A Story
165	A Most Unlikely Friendship
96	A Mother's Love
239	A Not So Rock-Solid Moon
41	A Real Bat Computer
30	A Real Sea Dragon
29	A Surprise Platypus
97	A Warm-Blooded Turtle
255	A Whale of a Song
134	After Their Kinds
178	Amazing Water
237	Amoeba Society
57	An Airplane from Ancient Egypt
157	An Alga that Flexes Its Muscles
223	An Important Extra Bone
32	Animal Culture
28	Animal Talk
104	Ant Antics!
108	Ant Mathematics
162	Ants Who Garden
167	Ants Who Live with Company
185	Are Birth Defects God's Plan?
139	Are Dinosaurs a Giant Mystery?
122	Are There Black Holes?
115	Aspirin
218	Astonishing "Coincidences"
258	Better Flower Power
118	Birds Who Build Pyramids
193	Bodyguard Ants
24	Brain Talk
103	Breaking Dollo's Law
142	Can There be Life Without God?
232	Caterpillar's Diet Choices
183	Cave Mysteries
249	Chimps Discover New Antibiotic!
151	Clean the Blushing Fish
254	Colorful Life in the Goldfish Bowl
83	Could Creationism Correct Science?
99	Could Leviathon Be a Dinosaur?
176	Could Science Make an Eye?
148	Creation and the Gospel
35	Creation Makes Better Science
74	Darwin's Child Murdered!

186	Darwin's Puzzle
203	Deceitful Orchids
50	Did Job Have a Weather Satellite?
208	Did the Sun Really Stand Still?
129	Do the Bible and Science Mix?
138	Does Genesis 2 Contradict Genesis 1?
68	Does the Lion's Tooth Bite Your Lawn?
53	Eating Pollution
241	Economic Life Among Bumblebees
114	Electric Avalanch
65	Electric Skin
105	Engineering Joint Lubrication
124	Engines of the Body
101	Even Bacteria Get Sick
146	Evolutionists Love "Lucy"
43	Evolution's Impact on Society
242	Facts that Weren't
11	Feline Secrets
5	Fish Learn in Schools, Too
205	Fish Teach Humans to Make Better Ceramics
92	Fishy Physicists
136	For the Sheer Joy of Variety!
152	Fungus that Blows Its Top
228	Genetic Music
201	Georgia's Grand Canyon
17	Giraffes in Antigravity Suits
37	God Protects Trees from Insects
132	God Shows Us the Earth From Space
89	Growing New Brain Cells
222	Harvesting Termites
174	Have Evolutionists Found a Bad Design in the Eye?
252	He Breathes Through His Legs
212	Hibernation: Not Simply Sleep
243	Horned Animals in Australia?
26	How Are Humans Different from Animals?
182	How Deep Is the Moon's Dust?
59	How Does the Nose Work?
180	How Old Does the Bible Say the Earth Is?
16	How to Keep Your Castle Fresh
214	How to Speak Crow
137	Humanity – Past, Present and Future
133	In the Midst of the Waters
79	Infants' Talk Puzzles Scientists
215	Instant Fish
72	Inviting Ants
169	Is Evolution Simply Change?
220	Is the Bible a Book of Medicine?
85	Is the Shark Related to the Pig?
189	James Clerk Maxwell
45	Jet Planes that Taste Bad
78	"Just So" Stories of Evolution
48	Katydids with Personal Guards
225	Language Studies Tell Us About Ourselves
253	Legs Knocked Out from Under Snake Evolution
91	Less Bang for the Red Shift
25	Life in Rock
31	"Liquid Air" Mimics God's System
199	Literate Butterflies

119	Love with Eight Arms
84	Magnetic Birds?
49	Mice Who Farm
172	Micro-Marvels of the Human Eye
195	Microscopic Bears
23	Millions of Noses
116	Mindless Logic?
147	Morality Is for Humans
190	More Butterfly Colors than You Can See
107	Natural Acid Rain
51	Natural Insect Repellent
211	Nature's Chemical Wars
10	Nature's Shark Repellent
213	New Discoveries About that Terrible Lizard
127	New Technique Supports Creation
177	One Designer
238	One Foot and a Sail
217	One Smart Woodpecker
198	Paleontologist Ants
230	Peanut Butter and Green Diamonds
94	Pet Your Houseplants
246	Phony Fungus Flowers
86	Plant Mathematicians
235	Plants Focus Light
216	Plastic Ceramics
66	Population and the Age of the Earth
187	Robert Boyle: Creation Scientist
73	Science and Miracles
100	Science Looks at Astrology
76	Science Sheds Light on the Darkest Day
181	Scientific Dating
207	Scientists Learn from Flies
81	Seeing Colors
144	Should the Sun Spin Faster?
231	Shrimp Kidnaps Chemist!
188	Sir Isaac Newton
126	Smart Heart
18	Social Life Among Lobsters
256	Stealthy Ants
191	Surprise of the Fire Salamander
13	Tasteful Data Preprocessing
102	Tent-Building Bats
120	The Aardwolf
149	The Amazing Woodpecker
9	The Animal that Confused Scientists
106	The Bare Bone Facts
123	The Bats Who Feed Trees
153	The Bat's Special Radar Design
209	The Bible Speaks About Our Ecological Crisis
60	The Big Universe
248	The Bird that Lies and Steals
250	The Birth that Ended Death
202	The Body's Amazing Ability to Adapt
113	The Body's Fleeting Workers
56	The Brain Repairs Itself
163	The Cowboy Lasso Mold
64	The Crafty Flea
69	The Creator Rebuilds Lives

164	The Cyanide Defense
131	The Days in Genesis
121	The Deep Diving Leatherback
75	The Destructive Power of Water
175	The Eye's Computer
20	The Facts of Human Life
171	The Faith of the Evolutionist
22	The Fish that Digs a Well
34	The Fossils Show Creation
210	The Games Animals Play
40	The Gift of Words
154	The Giraffe's Wonder Net
98	The Great Wall in Space
155	The Harvesting Ant
197	The Heart Speaks to the Brain
150	The Hi-Fi Cricket
166	The Hunting Wasp
173	The Incredibly Sensitive Eye
259	The Jellyfish with 24 Eyes
159	The Lizard with Hair
27	The Loving Lobster
7	The Loving Poison Dart Frog
179	The Marvel of Life
219	The Miracle Mother
194	The Miracle of New Life
90	The Miracle of Photosynthesis
77	The Miraculous Seed
161	The Most Improbable Honeybee
8	The Mountain of the Mists
42	The Mystery of the Frogfish
12	The Oldest Dinosaur
206	The Origin of Modern Language
221	The Power of God's Power
224	The Promise of Poison Frogs
33	The Punctual Bitterroot
158	The Sawfly
257	The Strangest Genetic Relationship
38	The Sun Praises Its Creator
135	The Sun, Moon and Stars
93	The Surprising Clown Fish
54	The Three-Eyed Porpoise
70	The Ultimate Engineer
19	The Universal Seven Day Week
87	The Universe's Missing Link
200	The Vanishing River
128	The Very Hairs of Your Head
196	The Walls Did Come Tumbling Down
245	The Warm-Blooded Bumblebee
36	The Wombat's Backward Pouch
110	The Wonders of Everyday Materials
140	The Work of a Superior Intelligence
204	The World's Most Amazing Bears
141	The World's Smallest Computer
21	The World's Strangest Bird
143	There Is No Simple Life
192	They Talk to the Trees
247	Thomas Jefferson Speaks Out
160	Time to Clean Fish

44	To Bee or Not to Bee
236	Traces of What Man Once Was?
80	U.S. Army Celebrates Its 75,000 Birthday!
95	Was Behemoth a Dinosaur?
145	We See By Faith
88	Well-Designed Snails
184	Were You Once a Fish?
15	Wernher von Braun on Creation
125	Whales Write Songs
240	Whales: Armed and Dangerous
170	What Is Faith?
117	When Facts Aren't Facts
112	When It's Better to Be Male
71	Where Is the Garden of Eden?
168	Who Is Against Evolution?
130	Who Is God?
6	Why Are There Germs?
234	Why Does it Rain?
46	Why Does the Sun Shine?
14	Why Don't You Rust?
111	Why the Boomerang Returns
109	Yes, Early Humans Wrote
244	Yoho Surprise!
229	Your 20-Watt Brain
62	Your Body's 100,000 Sentries

Contact us to order:

- "Letting God Create Your Day" books.

- "Creation Moments" CDs
  (each containing 30 programs).

- Ministry information and a free resource catalog.

    Creation Moments, Inc.
    P.O. Box 839
    Foley, MN 56329
    1-800-422-4253
    www.creationmoments.com